*Teratomas
and
Differentiation*

Teratomas and Differentiation

A Symposium
Held at the Roche Institute of Molecular Biology
Nutley, New Jersey, May 19-21, 1975

ORGANIZERS:

MICHAEL I. SHERMAN

Roche Institute of Molecular Biology
Nutley, New Jersey

ARNOLD J. LEVINE

Department of Biochemical Sciences
Princeton University
Princeton, New Jersey

PAUL BARTL

Roche Institute of Molecular Biology
Nutley, New Jersey

Teratomas and Differentiation

edited by

Michael I. Sherman
Roche Institute of Molecular Biology
Nutley, New Jersey

Davor Solter
The Wistar Institute of Anatomy and Biology
Philadelphia, Pennsylvania

Academic Press, Inc.   New York San Francisco London 1975
A Subsidiary of Harcourt Brace Jovanovich, Publishers

Academic Press Rapid Manuscript Reproduction

COPYRIGHT © 1975, BY ACADEMIC PRESS, INC.
ALL RIGHTS RESERVED.
NO PART OF THIS PUBLICATION MAY BE REPRODUCED OR
TRANSMITTED IN ANY FORM OR BY ANY MEANS, ELECTRONIC
OR MECHANICAL, INCLUDING PHOTOCOPY, RECORDING, OR ANY
INFORMATION STORAGE AND RETRIEVAL SYSTEM, WITHOUT
PERMISSION IN WRITING FROM THE PUBLISHER.

ACADEMIC PRESS, INC.
111 Fifth Avenue, New York, New York 10003

United Kingdom Edition published by
ACADEMIC PRESS, INC. (LONDON) LTD.
24/28 Oval Road, London NW1

Library of Congress Cataloging in Publication Data

Main entry under title:

Teratomas and differentiation.

  Bibliography: p.
  Includes index.
  1. Teratoma–Congresses. 2. Cell differentiation–Congresses. I. Sherman, Michael I.
II. Solter, Davor. III. Roche Institute of Molecular Biology.
RC254.5.T47 616.9'92 75-30900
ISBN 0–12–638550–5

PRINTED IN THE UNITED STATES OF AMERICA

This volume is dedicated to the memory of
GORDON TOMKINS

Contents

List of Contributors and Participants	xi
Preface	xv
Photographs of Contributors and Participants	xvi

I. INTRODUCTION AND TERMINOLOGY

Teratocarcinoma: Introduction and Perspectives   3
    G. Barry Pierce

Teratomas: Definitions and Terminology   13
    Leroy C. Stevens and G. Barry Pierce

II. EMBRYO-TERATOMA RELATIONSHIPS

Comparative Development of Normal and Parthenogenetic Mouse Embryos, Early Testicular and Ovarian Teratomas, and Embryoid Bodies   17
    Leroy C. Stevens

Teratomas from Haploid and Diploid Parthenogenetic Mouse Embryos   33
    C. F. Graham, M. W. McBurney, and S. A. Iles

Can Teratocarcinoma Cells Colonize the Mouse Embryo?   51
    R. L. Brinster

Developmental and Experimental Potentialities of Mouse Teratocarcinoma Cells from Embryoid Body Cores   59
    Beatrice Mintz, Karl Illmensee, and John D. Gearhart

Experimental Teratomas in Rats   83
    Nikola Skreb and Anton Svajger

III. SURFACE ANTIGENS ON EMBRYOS AND TERATOMAS

Expression of a Cell Surface Antigen Common to Primitive Mouse Teratocarcinoma Cells and Cleavage Embryos during Embryogenesis and Spermatogenesis   101
    Charles Babinet, Hubert Condamine, Marc Fellous, Gabriel Gachelin, Rolf Kemler, and Francois Jacob

Teratoma-Defined and Transplantation Antigens in Early Mouse
Embryos   109
    Michael Edidin and Linda Gooding

Surface Antigens of Blastocyst-Derived Cell Lines   123
    Richard A. Miller, Frank H. Ruddle, and Michael I. Sherman

IV. TERATOMA–HOST INTERACTIONS

Control of Teratocarcinogenesis   139
    Davor Solter, Nancy Adams, Ivan Damjanov, and Hilary Koprowski

Mutagenized Clones of a Pluripotent Teratoma Cell Line: Variants
with Decreased Differentiation or Tumour-Formation Ability   161
    Thierry Boon, Odile Kellerman, Elisabeth Mathy, and Jean A. Gaillard

V. CONTROL OF DIFFERENTIATION OF EMBRYONAL CARCINOMA CELLS

The Formation of Embryoid Bodies *In Vitro* by Homogeneous Embryonal
Carcinoma Cell Cultures Derived from Isolated Single Cells   169
    Gail R. Martin and Martin J. Evans

Differentiation of Teratoma Cell Line PCC4:azal *In Vitro*   189
    Michael I. Sherman

VI. PROPERTIES OF EMBRYONAL CARCINOMA CELLS

Ultrastructure of Murine Teratocarcinomas   209
    Ivan Damjanov and Davor Solter

Chromatin Properties of Mouse Primitive Teratocarcinoma Cells:
Electrophoretic Comparison with Somatic Tissues of the Mouse   221
    Jacques Jami and Jacques E. Loeb

VII. PROPERTIES OF TERATOMAS *IN VITRO*

The Differentiation of Clonal Teratocarcinoma Cell Cultures
In Vitro   237
    Martin J. Evans and Gail R. Martin

The *In Vitro* Differentiation of Embryoid Bodies Produced by a
Transplantable Teratoma of Mice   251
    J. D. Hall, M. Marsden, D. Rifkin, A. K. Teresky, and A. J. Levine

Alkaline Phosphatase Activity in Embryonal Carcinoma and Its
Hybrids with Neuroblastoma   271
    Edward G. Bernstine and Boris Ephrussi

CONTENTS

The Response of Murine Teratocarcinoma Cells to Infection with DNA
and RNA Viruses 289
    John M. Lehman, Iris B. Klein, and Rose M. Hackenberg

VIII. BIBLIOGRAPHY 305

Subject Index 319

List of Contributors and Participants

Asterisks denote contributors to this volume.

Joan Abbott, Memorial Sloan-Kettering Cancer Center, New York, New York 10021

Margaret Arny, Memorial Sloan-Kettering Cancer Center, New York, New York 10021

Karen Artzt (Session Chairwoman), Department of Anatomy, Cornell University Medical College, New York, New York 10021

Sui Bi Atienza, Roche Institute of Molecular Biology, Nutley, New Jersey 07110

*Charles Babinet, Department de Biologie Moleculaire, Institut Pasteur, 75 Paris 15e, France

Paul Bartl (Co-organizer), Roche Institute of Molecular Biology, Nutley, New Jersey 07110

Dorothea Bennett, Department of Anatomy, Cornell University Medical College, New York, New York 10021

*Edward Bernstine, Mammalian Genetics Section, Biology Division, Oak Ridge National Laboratory, Oak Ridge, Tennessee 37830

Arthur Blume, Roche Institute of Molecular Biology, Nutley, New Jersey 07110

*Thierry Boon, Department de Biologie Moleculaire, Institut Pasteur, 75 Paris 15e, France

*Ralph Brinster, Laboratory of Reproductive Physiology, School of Veterinary Medicine, University of Pennsylvania, Philadelphia, Pennsylvania 19103

Jerome Cantor, Roche Institute of Molecular Biology, Nutley, New Jersey 07110

Mario Capecchi, Department of Biology, University of Utah, Salt Lake City, Utah 84112

Christiane Crepin, Department of Genetics, University of California—San Francisco, San Francisco, California 94112

LIST OF CONTRIBUTORS AND PARTICIPANTS

Claire Cronmiller, Institute for Cancer Research, Fox Chase, Philadelphia, Pennsylvania 19111

*Ivan Damjanov, University of Connecticut Health Center, Farmington, Connecticut 06032

Michael Dewey, Institute for Cancer Research, Fox Chase, Philadelphia, Pennsylvania 19111

*Michael Edidin, Department of Biology, The Johns Hopkins University, Baltimore, Maryland 21218

Robert Erickson, Department of Pediatrics, University of California–San Francisco, California 94122

*Martin Evans, Department of Anatomy and Embryology, University College London, London WCIE 6BT, England

L. Patrick Gage, Roche Institute of Molecular Biology, Nutley, New Jersey 07110

*John Gearheart, Institute for Cancer Research, Fox Chase, Philadelphia, Pennsylvania 19111

*Linda Gooding, Division of Immunology, Duke Medical Center, Durham, North Carolina 27710

*Christopher F. Graham, Department of Zoology, University of Oxford, Oxford OX1 3PS, England

*Jennifer D. Hall, Department of Biochemical Sciences, Princeton University, Princeton, New Jersey 08540

Richard Hoar, Department of Pathology, Hoffmann-La Roche, Inc., Nutley, New Jersey 07110

*Karl Illmensee, Institute for Cancer Research, Fox Chase, Philadelphia, Pennsylvania 19111

Robert Jack, Institut für Genetik der Universitat zu Köln, 5 Köln 41, West Germany

*Jacques Jami, Institut de Recherches Scientifiques sur le Cancer du CNRS, 94800 Villejuif, France

Rudolf Jaenisch, Armand Hammer Center for Cancer Biology, Salk Institute for Biological Studies, San Diego, California 92112

Michael A. Jewett, Memorial-Sloan–Kettering Cancer Center, New York, New York 10021

Brenda Kahan, Department of Genetics, College of Agriculture and Life

LIST OF CONTRIBUTORS AND PARTICIPANTS

Sciences, University of Wisconsin, Madison, Wisconsin 53706

Barbara Knowles, The Wistar Institute, Philadelphia, Pennsylvania 19104

Gebhard Koch, Roche Institute of Molecular Biology, Nutley, New Jersey 07110

Les Kozak, The Jackson Laboratory, Bar Harbor, Maine 04609

Mary Ledbetter, Department of Microbiology, Dartmouth Medical School, Hanover, New Hampshire 03755

*John Lehman, Department of Pathology, University of Colorado Medical Center, Denver, Colorado 80220

*Arnold Levine (Co-organizer and Session Chairman), Department of Biochemical Sciences, Princeton University, Princeton, New Jersey 08540

Elwood Linney, Department of Biochemistry and Biophysics, University of California—San Francisco, San Francisco, California 94122

Martin Lubin, Department of Microbiology, Dartmouth Medical School, Hanover, New Hampshire 03755

*Gail Martin, Department of Anatomy and Embryology, University College of London, London WCIE 6BT, England

*Richard Miller, Department of Biology, Yale University, New Haven, Connecticut 06520

*Beatrice Mintz, The Institute for Cancer Research, Fox Chase, Philadelphia, Pennsylvania 19111

Leila Moustafa, National Institute of Environmental Health Sciences, Research Triangle Park, North Carolina 27709

Wade Parks, National Cancer Institute, c/o Meloy Laboratories, Rockville, Maryland 20850

*G. Barry Pierce (Session Chairman), Department of Pathology, University of Colorado Medical Center, Denver, Colorado 80220

Daniel Rifkin, The Rockefeller University, New York, New York 10021

*Frank Ruddle, Department of Biology, Yale University, New Haven, Connecticut 06520

Urs Rutishauser, The Rockefeller University, New York, New York 10021

David Salomon, Roche Institute of Molecular Biology, Nutley, New Jersey 07110

LIST OF CONTRIBUTORS AND PARTICIPANTS

Joe Sambrook, Cold Spring Harbor Biological Laboratories, Cold Spring Harbor, New York 11724

Melitta Schachner, Department of Neurosciences, Children's Hospital Medical Center, Boston, Massachusetts 02115

*Michael I. Sherman (Co-organizer and Session Chairman), Roche Institute of Molecular Biology, Nutley, New Jersey 07110

*Nikola Skreb, Department of Biology, Medical Faculty, University of Zagreb, Zagreb, Yugoslavia

*Davor Solter (Session Chairman), The Wistar Institute, Philadelphia, Pennsylvania 19104

*Leroy Stevens (Session Chairman), The Jackson Laboratory, Bar Harbor, Maine 04609

Richard Swarm, Department of Pathology, Hoffmann–La Roche, Inc., Nutley, New Jersey 07110

Jean-Paul Thiery, The Rockefeller University, New York, New York 10021

William C. Topp, Cold Spring Harbor Laboratory, Cold Spring Harbor, New York 11724

Arthur Weissbach, Roche Institute of Molecular Biology, Nutley, New Jersey 07110

Linda Wudl, Roche Institute of Molecular Biology, Nutley, New Jersey 07110

Preface

Each year, the Roche Institute of Molecular Biology holds a small and intensive symposium dealing with contemporary biomedical topics. This volume contains the proceedings of the fourth of these symposia, entitled "Teratomas and Differentiation." Although teratomas and teratocarcinomas are relatively rare in humans, these tumors develop spontaneously at high frequencies in some mouse strains, and can be easily induced in others. Advances with experimental teratomas and teratocarcinomas in mice over the last 15 years, and particularly in the last 5 years, have suggested a great potential for these tumors as model systems for the study of differentiation in oncology and embryology. As this potential has been recognized, the number of investigators in the field has climbed sharply with the inevitable result being a proliferation of new and exciting developments. In the face of this acceleration in research on experimental teratomas, the time appeared to be particularly appropriate for this, the first volume devoted entirely to the subject.

We wish to thank the Roche Institute of Molecular Biology for sponsoring this symposium. We are also grateful to Mrs. Denise Florkiewicz for helping with organizational arrangements and Mrs. Patricia Perkowski and Miss Linda Gregg for typing the manuscripts.

Dr. Gordon Tomkins was originally scheduled to participate in this Symposium. Dr. Tomkins died on July 22, 1975, following brain surgery. It is in recognition of his many important studies on the control of cell differentiation that we dedicate this volume to his memory.

N. Skreb, J. Abbott

A. Levine, L. Gooding, E. Bernstine

T. Boon, K. Artzt

L. Moustafa, E. Linney, R. Jack, M. Sherman

D. Solter, D. Bennett

M. Edidin

C. Graham, L. Stevens

F. Ruddle, E. Linney, K. Artzt, T. Boon

B. Pierce, F. Ruddle, R. Miller

C. Crepin, E. Bernstine, M. Sherman, J. Jami

C. Babinet

J. Abbott, B. Kahan, U. Rutishauser

Photographs by R. Welborn

I.
Introduction and Terminology

TERATOCARCINOMA: INTRODUCTION AND PERSPECTIVES

G. BARRY PIERCE

Department of Pathology
University of Colorado Medical Center
Denver, Colorado   80220

It is a great honor to be chosen to introduce the first conference in research on teratocarcinomas. The field has burgeoned at a rapid rate, a point of great satisfaction to Roy Stevens and me. In the late 1950's and early 1960's, either he or I formed a rather small, but intensely interested audience whenever the other gave a paper. To most oncologists, we were the fellows with 'the funny little tumors'. Obviously, funny little tumors have captured your interests too, and enough has been accomplished in this field so that perspectives may be drawn.

Through studies of strain 129 teratocarcinomas, the clinical classification of germinal tumors has been put on a sound basis (Stevens and Little, 1954; Stevens, 1959, 1960; Pierce and Dixon, 1959; Pierce et al., 1960). The observation that malignant stem cells of teratocarcinoma differentiate to benign somatic tissues forming a caricature of a normal process disproved the dogma 'once a cancer cell, always a cancer cell' (Pierce, et al., 1960). The discovery of the origin of embryonal carcinoma in primordial germ cells in strain 129 mice (Stevens, 1964, 1967) has led to an understanding of the role of normal stem cells in carcinogenesis. It turns out that all of this information extrapolates to other malignancies (Pierce and Wallace, 1971; Wylie et al., 1973). Tumors have their origin in normal stem cells, and give rise to a caricature of the renewal system of the particular tissue. Teratocarcinomas are thus in the main flow of tumor biology and their unique features promise more vital information. Finally, the use of teratocarcinomas as probes has great promise for the understanding of early development (Artzt et al., 1974). Let us consider these points briefly.

Figure 1 is the working classification of germinal tumors of the testis commonly used today. It was proposed by Dixon and Moore (1953), on the basis of their clinical studies of testicular tumors that occurred in the American Armed Forces in World War II, and is slightly modified by the work of Teilum (1959). The concept offered in the figure was based on morphological data, but it had developmental and functional implications which confounded many people

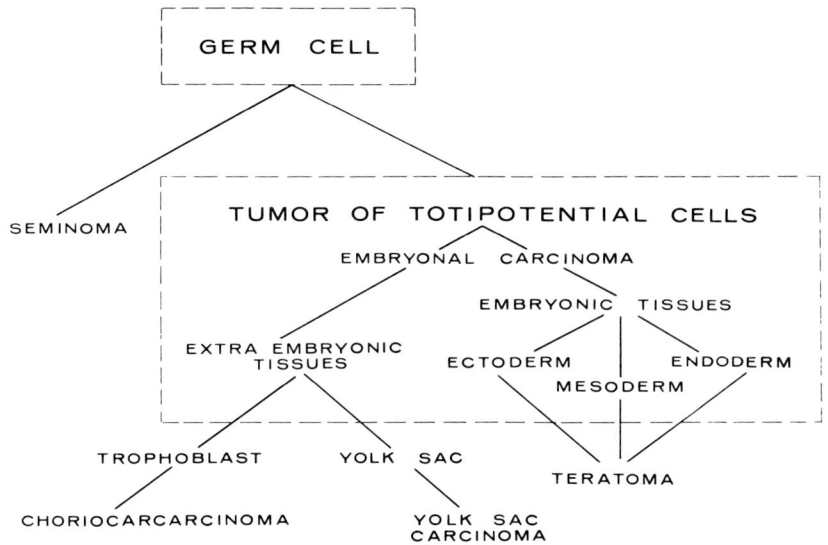

Figure 1. *Tumors of germ cell origin. From the classification of Dixon and Moore, 1953 and from Teilum, 1959.*

and, as a consequence, this classification was severely criticized. Briefly, the concept stated that embryonal carcinoma was a multipotential stem cell derived from primordial germ cells. If it differentiated into extraembryonic tissues, it could give rise to choriocarcinoma or yolk sac carcinoma. If it differentiated into embryonic tissues, representing each of the three germ layers, either a benign tera-

---

ABBREVIATIONS AND TERMINOLOGY

HCG - human chorionic gonadotropin.

toma or a teratocarcinoma would result. A teratocarcinoma is a malignant tumor which may be composed of derivatives of all three germ layers and which is malignant by virtue of the embryonal carcinoma cells which it contains. Seminoma, the most common of the germinal tumors, was believed to be unrelated to embryonal carcinoma. Many pathologists, particularly those in Europe, believed that seminoma was the primary tumor derived from germ cells which then gave rise to embryonal carcinoma.

In an effort to examine the validity of this classification, I heterotransplanted human testicular tumors and then studied them *in vitro* and with the electron microscope. Seminoma was resistant to transplantation or growth in tissue culture; by electron microscopic examination it was composed of a heterogeneous collection of cells. There were undifferentiated cells on the one hand and well differentiated elements on the other (Pierce, 1966). The latter had coils of mitochondria, reminiscent of mid piece of sperm, proacrosomal granules and intercellular bridges peculiar to spermatids (Rosai et al., 1969). It was concluded that seminoma was a caricature of spermatogenesis and its cells lacked multipotency. It was probably derived from spermatocytes, not directly from primordial germ cells or spermatogonia and it was not directly related to embryonal carcinoma. Studies on embryonal carcinoma confirmed this conclusion (Pierce et al., 1960; Pierce, 1966). Four human embryonal carcinomas and four choriocarcinomas were successfully heterotransplanted (Pierce et al., 1958; Verney and Pierce, 1959). If an embryonal carcinoma contained somatic tissues, it would not grow in an immunologically abrogated heterologous host. The first embryonal carcinoma to be successfully heterotransplanted synthesized human chorionic gonadotropin (HCG) in 20% of the animals in which it was grafted as evidenced by the genital response of the host. This observation led to the idea that this embryonal carcinoma, which lacked the morphological appearance of choriocarcinoma, was probably spontaneously differentiating toward choriocarcinoma, and in 20% of cases had commenced synthesizing small amounts of HCG.

Choriocarcinoma is composed of masses of cytotrophoblast overlain by multinucleated syncytiotrophoblastic giant cells. This tumor grew well in heterologous hosts, although the orientation of cyto- and syncytiotrophoblast was usually disturbed. The tumors always synthesized human chorionic gonadotropin (Verney and Pierce, 1959). By immunohistochemistry, and later by electron microscopy, it was demonstrated that the syncytium was exceedingly well differentiated (Pierce and Midgley, 1963) and synthesized HCG (Midgley and Pierce, 1962). If syncytiotrophoblast was differentiated and the cytotrophoblast was undifferentiated, were they developmentally related? Using normal monkey placenta it was shown that cytotrophoblast was labeled 2 hours after intravenous injection of tritiated thymidine but the syncytium was not labeled. At 48 hours, label was found in the syncytium (Midgley et al., 1963). Later, by electron microscopy, it was determined that cytotrophoblastic cells were incorporated into the syncytium through breaks in the plasma membranes of the cells (Pierce et al., 1964). Thus, it was concluded that embryonal carcinoma differentiated into cytotrophoblast which could only synthesize small quantities of HCG, and in turn the cytotrophoblast differentiated into the syncytium which was the differentiated form of trophoblast and synthesized the bulk of the HCG. The net result was the production of a choriocarcinoma. Obviously this was not proof positive, but the experiments to be described later in teratomas of strain 129 mice solidified the notion.

Choriocarcinoma is not found in rodents and for a long time tumors of extraembryonic tissue were not recognized in rodents. Now malignant tumors that are identified as being derived from distal endoderm have been observed in the rat (Symeonidis and Mori-Charez, 1952), mouse (Nakahara et al., 1967) and hamster (H. Pogosianz, personal communication). These tumors synthesize large amounts of basement membrane, which is the counterpart of Reichert's membrane of the rodents' placenta. Since Reichert's membrane and distal endoderm form the parietal wall of the yolk sac, these tumors have been named parietal yolk sac carcinomas (Pierce et al., 1962). Visceral yolk sac carcinomas have not been identified as such. The parietal yolk sac carcinoma converts readily to the ascites and synthesizes basement membrane in the centers of aggregates of free-floating neoplastic cells. This has proved to be a model for the study of synthesis, biochemistry, response to injury and degradation of basement membrane. The basement membrane antigens contained in these cells are fine markers for identifying distal endoderm cells.

Distal endoderm cells can be isolated from strain 129 teratocarcinomas. They are often benign at the time of first isolation and resistant to growth *in vitro*, but many are malignant from the outset and grow well either subcutaneously, interperitoneally or in tissue culture. There seems to be a propensity for best growth in the interperitoneum or in tissue culture (Pierce and Dixon, 1959; Pierce and Verney, 1961). Through study of these tumors in mice it was possible to confirm Teilum's observation that a controversial tumor of testicles of small boys was a yolk sac carcinoma (Pierce et al., 1970).

Because of the difficulty in growing human teratocarcinomas in heterologous hosts, we studied the ovarian teratocarcinomas first described by Fekete and Ferrigno (1952) and then the testicular teratocarcinomas isolated by Stevens and Little (1954). When transplantable teratocarcinoma 402 VI A was converted to the ascites, embryoid bodies were produced (Pierce et al., 1960; Stevens, 1960). These varied in size from large cystic ones containing three germ layers to small solid ones containing two germ layers and visible only with the microscope. The cystic ones were employed in establishing the nature of teratocarcinoma. If transplanted in animals they were teratocarcinogenic, and in timed studies, it was established that the endoderm covering the embryoid body and mesenchyme were not tumorigenic (Pierce and Dixon, 1959; Pierce et al., 1960). The embryonal carcinoma invaded through the wall of the embryoid body and gave rise to new tumors. When these were less than a millimeter in diameter, they were composed exclusively of embryonal carcinoma, but when they become larger they contained elements representing each of the germ layers. It was concluded that embryonal carcinoma was the multipotent stem cell of these tumors. Stevens (1959) arrived at the same conclusions in his study of the development of primary tumors.

The very large cystic embryoid bodies that measured more than 6 mm in diameter often contained no embryonal carcinoma: it had either perished or differentiated. When these large cystic embryoid bodies were transplanted, it was possible to determine the biological behavior of tissues derived from embryonal carcinoma. These embryoid bodies gave rise to the equivalent of benign dermoid cysts of the ovary of women. The tissues contained in these dermoid cyst-like transplants did not grow for periods of 6 months and it was concluded that the embryonal carcinoma had in fact differentiated into benign cells in tissues (Pierce et al., 1960).

The mode of development of embryoid bodies was studied in the ascites and *in vitro* (Pierce et al., 1960; Pierce and Verney, 1961). It turned out that embryonal carcinoma of explants of teratocarcinomas budded from surfaces. Some of the surface cells differentiated into proximal endoderm which invested the cancer cells and formed an aggregate equivalent to a 5-6 day mouse embryo. Then some of the cancer cells beneath the endoderm differentiated into mesenchyme forming an embryoid body of 3 layers. The adult somatic tissues of the tumors then developed from the endoderm and mesenchyme. Brain and skin, and endoderm, of course, developed directly from the embryonal carcinoma. It was concluded that the teratocarcinoma was a caricature of embryogenesis.

The experiments of Kleinsmith and Pierce (1964) confirmed the multipotency of embryonal carcinoma. The cloning experiments were done *in vivo* since embryonal carcinoma cells grow as heaped-up colonies loosely attached to the substrate. We were afraid that if we cloned *in vitro* that embryonal carcinoma cells might move about, and we would never really be sure of the results of the supposed cloning experiments.

The observation that malignant cells differentiated into benign cells suggests attempts at enhancing and directing differentiation as an alternative to the usual types of cytotoxic chemotherapy. Obviously, if embryonal carcinoma is to be directed in its differentiation, the end point of the experiment must be either a yolk sac carcinoma, benign proximal endoderm, mesenchyme, nervous tissue or squamous epithelium. These are the only things into which embryonal carcinoma can differentiate. We geared our attacks to produce parietal yolk sac carcinoma because of its immunohistochemical marker, and the capacity for the cells to grow competitively with embryonal carcinoma in tissue culture. We have been unsuccessful in these experiments, but we have learned a bit about teratocarcinomas during their performance.

Low pH and low temperature in roller cultures seem to favor development of parietal yolk sac carcinoma from embryonal carcinoma (Pierce and Verney, 1961; Lehman et al., 1974).

If small cubes of teratocarcinoma are explanted, epithelium grows over the surface, effectively confining the explant as an organ culture; differentiated tissues die, leaving embryonal carcinoma and sometimes parietal yolk sac carcinoma embedded in necrotic debris (Pierce and Verney, 1961). If these cultures, devoid of mesenchyme, are transplanted back into mice, instead of masses of neural tissue, the resultant teratocarcinomas contain striated muscle, as much as 80% by volume. With second and third passages in mice, the picture reverts to the usual situation dominated by neural tissue. Obviously, the *in vivo* environment favors neurogenesis, the *in vitro* seems to select embryonal carcinoma for a propensity to form mesenchyme and muscle. Monolayer cultures do not behave in this fashion (Pierce and Verney, 1961; Rosenthal et al., 1970; Evans, 1972; Teresky et al., 1974).

Possibly a better example of directed differentiation

is the observation by Speers and Lehman that embryonal carcinoma grown on a substrate of parietal yolk sac differentiates into squamous epithelium (Speers and Lehman, in preparation). The time element is more manageable in these observations and may be amenable to molecular attack.

Direction of differentiation is where the action is. This is one of the few alternatives to cytotoxic chemotherapy available today. It turns out in studies of squamous cell carcinomas of the skin (Pierce and Wallace, 1971), breast carcinoma (Wylie et al., 1973) and neuroblastoma (Goldstein, 1972) that just as teratocarcinoma is a caricature of embryogenesis, these tumors are caricatures of tissue genesis. The differentiated elements in these tumors are post-mitotic and senescent. They are derived from malignant stem cells. We must learn how to direct the differentiation of the latter.

Let us now focus on Stevens' discovery that embryonal carcinoma has its origin in primordial germ cells (Stevens, 1964, 1967). It has been a dogma, although a disliked one, that malignant cells arise from differentiated cells by a process of dedifferentiation. Stevens showed that a tumor, albeit a funny little tumor, arose from the stem cells of the species. The normal and malignant stem cells had comparable ultrastructure (Pierce et al., 1967). This probably means that normal stem cells have alternative pathways of differentiation open to them, one recognizable as a malignant pathway. A cell, an ovum, placed in the uterus of a pregnant animal gives rise to a new animal. If placed in the testes or beneath the renal capsule it gives rise to a tumor. Why? Are the hormones of pregnancy important? Have the transplanters ever placed an ovum beneath the renal capsule of a pregnant animal?

Does Stevens' observation of the origin of a tumor and a stem cell extrapolate to other neoplastic systems? The answer is 'of course.' The stem cells of lactiferous epithelium are contained in the simple ducts observed in the resting breast between pregnancies. With the stimulation of pregnancy and nursing there is proliferation and differentiation of progeny of these cells into lactating cells. If the ultrastructure of stem cells of normal breast is compared to that of neoplastic mammary stem cells, it becomes apparent that they are equally undifferentiated or differentiated, as one chooses to interpret the data. In the former there is control which evolves tissue which we recognize as normal, and in the latter the controls evolve clinical malignancy. The point to be made is that the stem cells of normal breast and in normal gut are in the same degree of differentiation as are their malignant counterparts. This is why I believe malignant stem cells are derived from normal ones.

In conclusion, no other funny little tumor has contributed so much to the understanding of the cell biology of malignancy. Because of their unique characteristics they promise to unravel not only more of the secrets of malignancy but important secrets of early development.

REFERENCES

Artzt, K., Bennett, D., Jacob, F. (1974). Primitive teratocarcinoma cells express a differentiation antigen specified by a gene at the T-locus in the mouse. Proc. Nat. Acad. Sci. U.S. 71, 811.
Dixon, F. J., Jr., Moore, R. A. (1953). Testicular tumors- A clinicopathologic study. Cancer 6, 427.
Evans, M. J. (1972). The isolation and properties of a clonal tissue culture strain of pluripotent mouse teratoma cells. J. Embryol. Exp. Morph. 28, 163.
Fekete, E., Ferrigno, M. A. (1952). Studies on a transplantable teratoma of the mouse. Cancer Res. 12, 438.
Goldstein, M. N. (1972). Growth and differentiation of normal and malignant sympathetic neurons *in vitro*. In: "Cell Differentiation" (Eds. R. Harris, P. Allin and D. Viza). Munksgaard, Copenhagen, p. 131.
Jami, J., Ritz, E. (1974). Multipotentiality of single cells of transplantable teratocarcinomas derived from mouse embryo grafts. J. Nat. Cancer Inst. 52, 1547.
Kleinsmith, L. J., Pierce, G. B. (1964). Multipotentiality of single embryonal carcinoma cells. Cancer Res. 24, 1544.
Lehman, J. M., Speers, W. C., Swartzendruber, D. E., Pierce, G. B. (1974). Neoplastic differentiation: Characteristics of cell lines derived from a murine teratocarcinoma. J. Cell Physiol. 84, 13.
Martin, G. R., Evans, M. J. (1974). The morphology and growth of a **pluripotent** teratocarcinoma cell line and its derivatives in tissue culture. Cell 2, 163.
Midgley, A. R., Pierce, G. B. (1962). Immunohistochemical localization of human chorionic gonadotropin. J. Exp. Med. 115, 289.
Midgley, A. R., Jr., Pierce, G. B., Jr., Deneau, G. A., Gosling, J. R. G. (1963). Morphogenesis of syncytiotrophoblast *in vivo* - An autoradiographic demonstration. Science 141, 349.
Nakahara, W., Tokuzen, R., Fukuoka, F. (1967). A trans-

plantable hyalinogenic tumor of the mouse. Gann 58, 475.

Pierce, G. B., Bullock, W. K., Huntington, R. W. (1970). Yolk sac tumors of testis. Cancer 25, 644.

Pierce, G. B., Jr., Dixon, F. J. Jr. (1959). Testicular teratomas. I. Demonstration of teratogenesis by metamorphosis of multipotential cells. Cancer 12, 573.

Pierce, G. B., Dixon, F. J. (1959). Testicular teratomas. II. A teratocarcinoma as an ascites tumor. Cancer 12, 584.

Pierce, G. B., Dixon, F. J., Verney, E. L. (1958). The biology of testicular cancer. II. Endocrinology of transplanted tumors. Cancer Res. 18, 204.

Pierce, G. B., Dixon, F. J., Jr., Verney, E. L. (1960). Teratocarcinogenic and tissue forming potentials of the cell types comprising neoplastic embryoid bodies. Lab. Invest. 9, 583.

Pierce, G. B., Midgley, A. R., Jr. (1963). The origin and function of human syncytiotrophoblastic giant cells. Amer. J. Path. 43, 153.

Pierce, G. B., Midgley, A. R., Jr., Beals, T. F. (1964). An ultrastructural study of the differentiation and maturation of trophoblast. Lab. Invest. 13, 451.

Pierce, G. B., Midgley, A. R., Feldman, J. D., Sri Ram, J. (1962). Parietal yolk sac carcinoma. Clue to the histogenesis of Reichert's membrane of the mouse embryo. Amer. J. Path. 41, 549.

Pierce, G. B., Stevens, L. C., Nakane, P. K. (1967). Ultrastructural analysis of the early stages of development of teratocarcinomas. J. Nat. Cancer Inst. 39, 755.

Pierce, G. B., Verney, E. L. (1961). An *in vitro* and *in vivo* study of differentiation in teratocarcinomas. Cancer 14, 1017.

Pierce, G. B., Wallace, C. (1971). Differentiation of malignant to benign cells. Cancer Res. 31, 127.

Rosai, J., Khodadoust, K., Silber, I. (1969). Spermatocytic seminoma. II. Ultrastructural study. Cancer 24, 103.

Rosenthal, M. D., Wishnow, R. M., Sato, G. H. (1970). *In vitro* growth and differentiation of clonal populations of multipotential mouse cells derived from a transplantable testicular teratocarcinoma. J. Nat. Cancer Inst. 44, 1001.

Stevens, L. C. (1959). Embryology of testicular teratomas in strain 129 mice. J. Nat. Cancer Inst. 23, 1249.

Stevens, L. C. (1960). Embryonic potency of embryoid bodies derived from a transplantable testicular teratoma of the mouse. Develop. Biol. 2, 285.

Stevens, L. C. (1964). Experimental production of testicular teratomas in mice. Proc. Nat. Acad. Sci. U. S. 52, 654.

Stevens, L. C. (1967). Origin of testicular teratomas from primoridal germ cells in mice. J. Nat. Cancer Inst. 38, 549.

Stevens, L. C., Little, C. C. (1954). Spontaneous testicular teratomas in an inbred strain of mice. Proc. Nat. Acad. Sci. U.S. 40, 1080.

Symeonidis, A., Mori-Chavez, P. (1952). Transplantable ovarian papillary adenocarcinoma of rat with ascites implants of the ovary. J. Nat. Cancer Inst. 13, 409.

Teilum, G. (1959). Endodermal sinus tumor of the ovary and testis. Cancer 12, 1092.

Teresky, A. K., Marsden, M., Kuff, E. L., Levine, A. J. (1974). Morphological criteria for the *in vitro* differentiation of embryoid bodies produced by a transplantable teratoma of mice. J. Cell Physiol. 84, 319.

Verney, E. L., Pierce, G. B. (1959). The biology of testicular tumors. III. Heterotransplanted choriocarcinoma. Cancer Res. 19, 633.

Wylie, C. V., Nakane, P. K., Pierce, G. B. (1973). Degree of differentiation in nonproliferating cells of mammary carcinoma. Differentiation 1, 11.

TERATOMAS: DEFINITIONS AND TERMINOLOGY

LEROY C. STEVENS[1] AND G. BARRY PIERCE[2]

[1]The Jackson Laboratory
Bar Harbor, Maine   04609
and [2]Department of Pathology
University of Colorado Medical Center
Denver, Colorado   80220

We are recommending nomenclature with the sole objective of preventing a proliferation of synonyms, even though present terminology may not be considered adequate. Until recently, there has not been much unanimity in terminology of germinal tumors used by clinicians, and the input from developmental studies with a proliferation of new terms could lead to even greater confusion.

Gonadal teratomas of mice are composed of tissues derived from all three germ layers, and they are usually benign (Fig. 1). They originate from germ cells, and during their early development are composed of cells that closely resemble inner cell mass cells of blastocysts or embryonic ectoderm at five or six days of gestation. These cells differentiate into many kinds of tissues. Some of these tumors are malignant and contain multipotential stem cells which have been named *embryonal carcinoma cells* because of their resemblance to embryonic cells. Tumors containing embryonal carcinoma cells and elements of all three germ layers have been named *teratocarcinomas* (Fig. 2). Embryonal carcinoma cells invade and may metastasize. If all of the embryonal carcinoma cells of a tumor differentiate, the tumor stops growing and is again referred to as a *teratoma*. On occasion, embryonal carcinoma cells may lose their capacity to differentiate and the resultant tumors are referred to as *embryonal carcinomas* (Fig. 3).

Sometimes embryonal carcinoma cells give rise to tumors of extraembryonic fetal membranes which consist exclusively

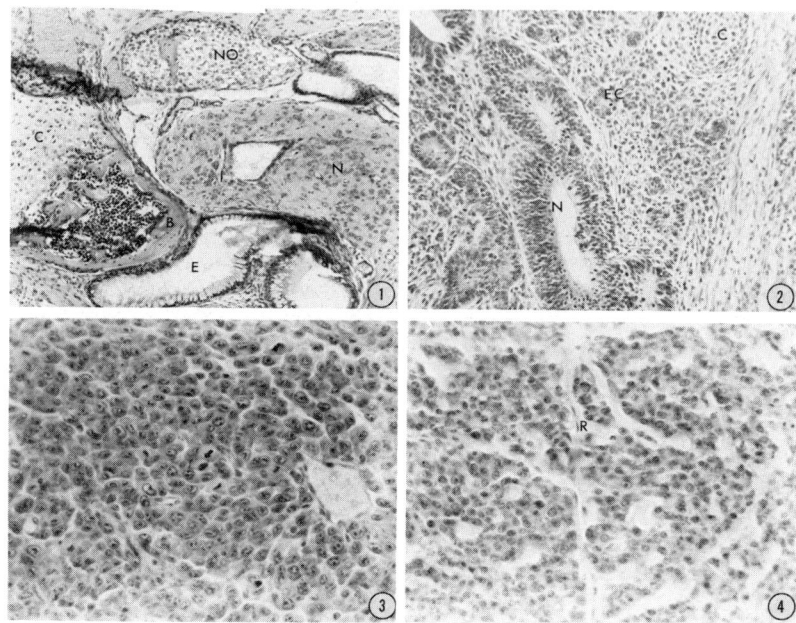

Figure 1. Teratoma composed of mature tissues from all three germ layers including notochord (NO), neural tissue (N), cartilage (C), bone (B) with marrow, and epithelial cysts (E).
Figure 2. Teratocarcinoma composed of embryonal carcinoma cells (EC), immature neuroepithelium (N), and cartilage (C).
Figure 3. Embryonal carcinoma lacking differentiated cells.
Figure 4. Parietal yolk sac carcinoma composed of parietal endoderm that secretes Reichert's membrane (R).

of distal endodermal cells that secrete Reichert's membrane. This appears to be an irreversible change and these tumors are referred to as *parietal yolk sac tumors* (Fig. 4). Some of these are apparently benign but others are malignant and may be referred to as *parietal yolk sac carcinomas*.

Transplantable teratocarcinomas may be experimentally produced by grafting early embryos to various sites in adults. These are indistinguishable from those derived from ovarian and testicular teratocarcinomas and the term *embryo derived teratocarcinoma* may be used to indicate their origin.

II.
Embryo-teratoma Relationships

COMPARATIVE DEVELOPMENT OF NORMAL AND PARTHENOGENETIC MOUSE EMBRYOS, EARLY TESTICULAR AND OVARIAN TERATOMAS, AND EMBRYOID BODIES

LEROY C. STEVENS

The Jackson Laboratory
Bar Harbor, Maine   04609

INTRODUCTION

Normal mouse embryos develop from eggs that have been fertilized by spermatozoa. Parthenotes develop from eggs that are ovulated and begin development without being fertilized (Tarkowski, 1970; Graham, 1970). Ovarian teratomas result from parthenogenetic activation of eggs within the ovaries of mice older than one month (Stevens, 1975). Testicular teratomas originate from primordial germ cells that begin to develop within the seminiferous tubules of fetuses during the twelfth day of gestation (Stevens, 1964, 1966, 1967a,b, 1970a; Pierce et al., 1967). Embryoid bodies are produced by testicular and ovarian teratomas (Pierce and Dixon, 1959a,b; Pierce, 1961; Stevens, 1959; Stevens and Varnum, unpublished results). All five of these developmental situations initially originate from germ cells, and they all have developmental characteristics in common. They appear similar in some respects, and different in others. This article will attempt to point out some of these similarities and differences.

MATERIALS AND METHODS

Normal embryos were derived from several inbred strains. Females were examined for copulation plugs, and the day plugs were found was considered as day-0.

Parthenogenetic embryos were obtained from inbred strain

LT/Sv mice. When large numbers of LT/Sv virgin mice are autopsied, about 10% of them have uterine swellings that contain parthenogenetic embryos. Usually these embryos survive for about a week, and then they are aborted.

Most pregnant virgins had only one or two parthenogenetic embryos. To increase the number of embryos per female, groups of mice were superovulated and mated with vasectomized males. On day-1 of gestation, two-cell eggs were flushed out of the oviduct and cultured to the blastocyst stage in Whitten's (1971) medium. Groups of about 9 blastocysts were transferred to the uteri of pseudopregnant hosts. Most of them implanted in the uterus and preliminary observations suggest that they developed better than in mice with only one or two embryos.

Ovarian teratomas were found in strain LT/Sv females. Most of them were detectable grossly, but some were microscopic and were observed during histological examination of the ovaries. Teratomas were not observed in mice younger than one month of age. At one month, the incidence is low. It is higher in two month old mice, and at three months it is about 50%. Early stages in development of teratomas were observed in three month old mice. The incidence of teratomas does not increase in mice older than three months for reasons that are not yet understood (Stevens and Varnum, unpublished).

Spontaneous testicular teratomas in early stages of development were found by histological examination of thousands of testes from 15- to 18-day strain 129 fetuses.

Experimentally induced testicular teratomas were produced by grafting 12- and 13-day genital ridges into the testes of adults. Grafted male genital ridges develop into testes and, for some strains of mice, most of them have several foci of teratomas that can be identified 6 days after grafting. The development of spontaneous and experimentally induced teratomas is identical.

Embryoid bodies were obtained from transplantable testicular and ovarian teratocarcinomas maintained in ascites form. Most teratomas in mice are benign. Usually all of the undifferentiated pluripotent stem cells differentiate into non-proliferating tissues. Occasionally, however, they are malignant, and can be established as transplantable tumors. When some transplantable teratocarcinomas are converted to ascites tumors, they form structures that resemble normal embryos in morphology and in embryonic potency (Stevens, 1960).

All tissues were fixed in Vandegrift's solution, sectioned at 6 microns, and stained in hematoxylin and eosin.

OBSERVATIONS

The stage of development of mouse embryos is usually expressed as the number of days of gestation. This would not have much meaning in a comparative study. Theiler (1972) classified mouse embryos into stages of development. I will use his classification with the equivalent to days of gestation of normal development in parentheses.

Stage 1 (0 to 20 hours). Parthenotes are ovulated as 1-cell eggs, as are normal eggs.

Stage 2 (20 to 40 hours). In normal and parthenogenetic development, 2-cell eggs were found in the oviduct on day-1 of gestation (Fig. 1). Two-cell eggs were also observed in all of the ovaries of mature LT/Sv mice examined (Fig. 2). This is regarded as an early stage in the development of a teratoma. Fragmented eggs are also very common in ovaries of strain LT/Sv mice.

Stage 4 (3 days). Morulae equivalent to 3 days of normal development (Figs. 3 and 4) were much less common in the ovaries of LT/Sv mice than were 2-cell eggs. Apparently most of the parthenogenetically activated ovarian eggs die during very early cleavage stages.

Parthenogenetic morulae were flushed from the uteri of females mated with vasectomized males. Most of them survived in culture to the blastocyst stage, indicating that they were viable.

Cells resembling those in stage 2 or 4 embryos were not seen in spontaneous or experimentally induced testicular teratomas or in any transplantable teratocarcinoma of ovarian or testicular origin.

Stage 5 and 6 (4 to 4 1/2 days). Parthenogenetic blastocysts equivalent to those floating free in the uteri of normal (Fig. 5) 4-day pregnant mice were flushed from the uteri of females mated with vasectomized males. Others were obtained by culturing 2-cell eggs for three days.

Blastocysts were rarely observed in the ovaries of LT/Sv mice. Mouse embryos develop very rapidly, and the probability of finding early embryos at a particular stage of development in an ovary is low. The ovarian blastocyst shown in Fig. 6 is at the stage of implantation. A layer of endodermal cells is recognizable on the surface of the inner cell mass. Theiler (1972) described an embryo at the same stage of development which had 243 cells.

Structures resembling blastocysts have not been observed in spontaneous, induced, or transplantable testicular teratomas. Transplantable ovarian teratocarcinomas are also devoid of blastocysts.

Stage 7 (5 days). By 5 days of gestation the inner cell mass of normal and parthenogenetic embryos has enlarged considerably and bulges into the cavity lined by trophectoderm. The ectoderm is enclosed by a single-cell layer of primary endoderm. The embryonic ectoderm can usually be distin-

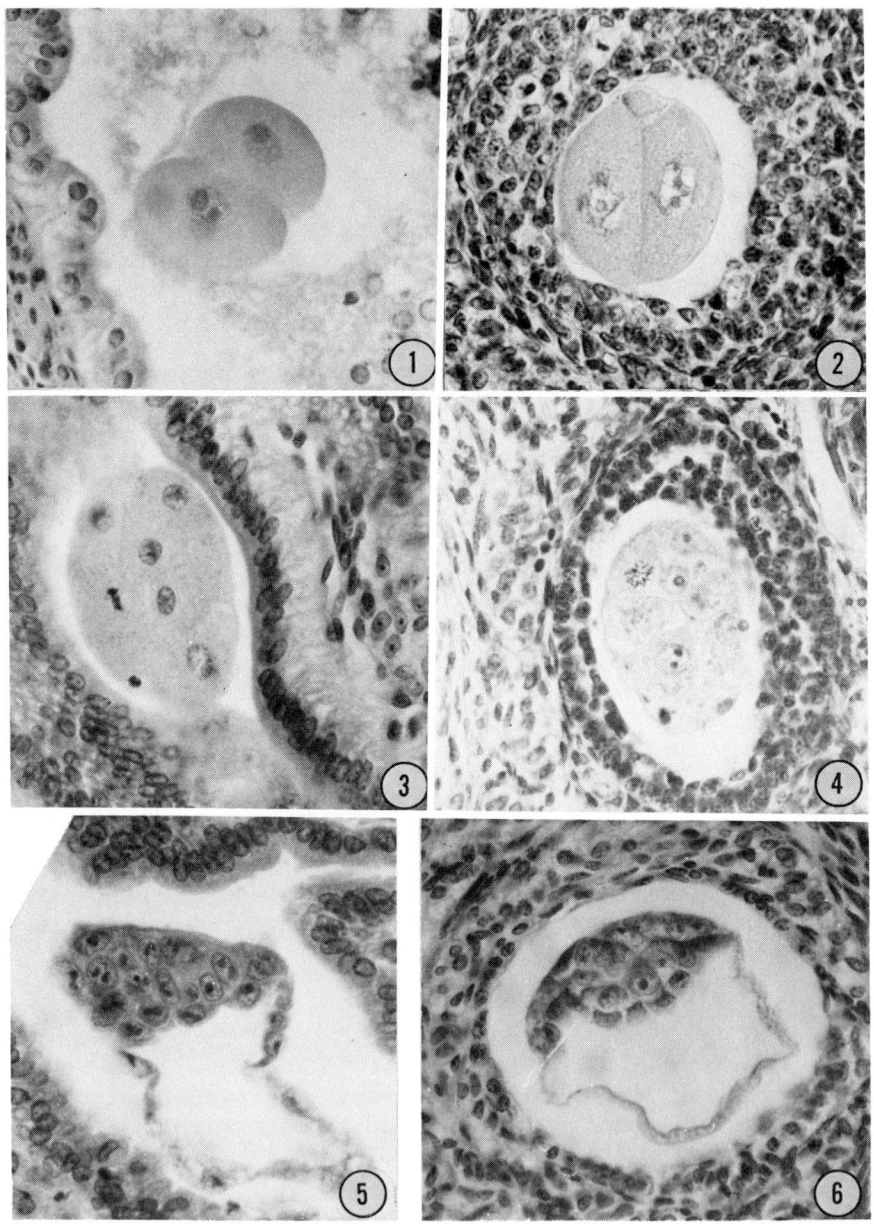

guished from the extra-embryonic ectoderm (Fig. 7). We have not observed this stage in ovarian embryos, but since earlier and later stages were found, we assume that early ovarian teratomas pass through this stage. Early testicular teratomas in 15- or 16-day festuses are composed of clumps of cells that resemble inner cell mass cells of normal blastocysts (Fig. 8). Early testicular teratomas differ from normal and parthenogenetic embryos, ovarian teratomas, and embryoid bodies (even those of testicular teratocarcinomas) in that they do not produce proximal or distal endoderm with Reichert's membrane.

Stage 8 (6 days). Normal embryos slightly older than that shown in Fig. 7 (5 days) develop a central cavity -- the proamniotic cavity. Cavities that are probably homologous to the proamniotic cavity are regularly seen in parthenogenetic embryos (Fig. 9), early testicular (Fig. 10) and ovarian teratomas (Fig. 11) and embryoid bodies (Fig. 12).

Stage 9 (6 1/2 days). The cells lining the proamniotic cavity elongate with their long axes perpendicular to it. This is illustrated for normal embryos in Fig. 13, parthenotes (Fig. 14), ovarian and testicular teratomas (Figs. 15 and 16), and embryoid bodies (Figs. 17, 19, and 25). The ectodermal cells in the ovarian teratoma illustrated in Fig. 15 are not elongated as in normal embryos, but the distinction between endoderm and ectoderm is clear. The cells of this embryo are large and probably polyploid.

Distal endoderm with Reichert's membrane and trophoblastic giant cells are regularly present in implanted parthenotes (Fig. 14) and ovarian embryos (Fig. 24), but not in spontaneous testicular teratomas. Transplantable testicular and ovarian teratocarcinomas frequently contain distal endoderm with Reichert's membrane. In fact, they often become limited to distal endoderm in their differentiative capacity. Some ascites transplantable teratocarcinomas have embryoid bodies composed of envelopes of distal endoderm filled with Reichert's membrane.

Stage 10 and 11 (7 to 7 1/2 days). At this stage, mesoderm is formed by proliferating cells in the primitive streak (Figs. 21 and 22). The mesodermal cells migrate anteriorly,

Figure 1. Normal two-cell egg in oviduct.
Figure 2. Parthenogenetic two-cell egg in ovarian follicle. This represents an early stage in development of a teratoma.
Figure 3. Normal morula in oviduct.
Figure 4. Parthenogenetic morula in ovarian follicle.
Figure 5. Normal blastocyst in uterus.
Figure 6. Parthenogenetic blastocyst in ovarian follicle.

posteriorly, and laterally between the ectoderm and endoderm. Parthenotes only rarely survive to this stage. Unusually advanced parthenotes are shown in Figs. 23 and 26. They were composed of trophoblastic giant cells, Reichert's membrane, distal endoderm, proximal endoderm, mesoderm, and ectoderm.

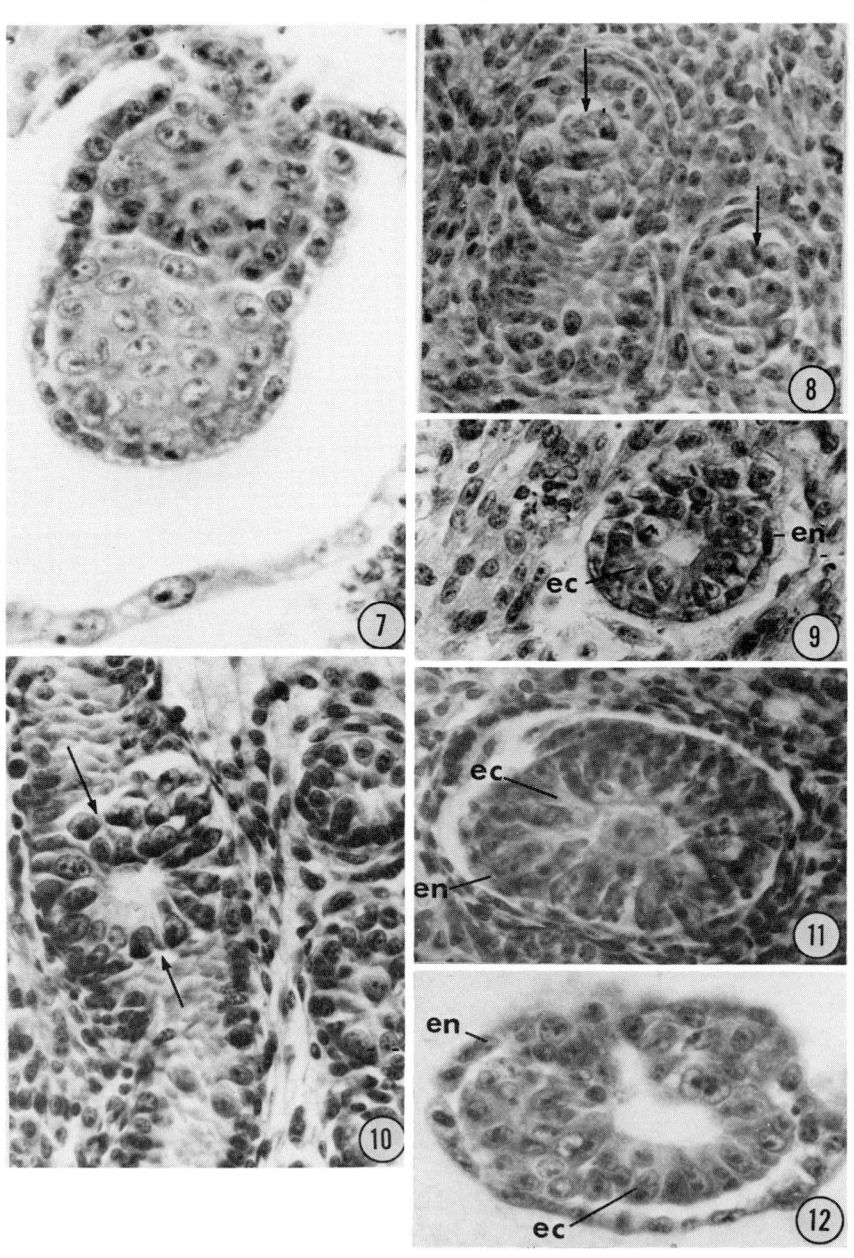

One of these parthenotes had an amniotic cavity enclosed by an abnormally thick amnion and by embryonic ectoderm (Fig. 26). Only two parthenotes more advanced than those shown in Figs. 23 and 26 have been observed thus far. One was very disorganized, but fetal heart muscle could be recognized. The other was dead and had about six pairs of somites, a wavy neural tube and heart muscle.

Ovarian embryos very rarely retain their embryonic architecture to the primitive streak stage (Fig. 20). An unusually well organized one is shown in Fig. 24. Its composition is similar to the parthenote shown in Fig. 23.

Embryoid bodies with areas resembling primitive streaks are common in some transplantable ascites teratocarcinomas (Fig. 25). The ectoderm is continuous with non-epithelial cells that have the same relationship to ectoderm and mesoderm as in normal embryos. All embryoid bodies have an outer layer of endoderm enclosing undifferentiated non-epithelial cells or epithelial cells resembling early ectoderm. Mesoderm formation does occur in embryoid bodies as indicated by the presence of blood islands with immature blood cells.

Spontaneous testicular teratomas have never been observed with embryoid structures of this stage.

When simple embryoid bodies that lack organized ectoderm are cultured *in vitro*, they increase in size, and form ectoderm and mesoderm (Hsu and Baskar, 1974). Simple embryoid bodies *in vivo* split apart and do not increase in size.

Testicular teratomas in mature mice resemble ovarian teratomas. They have the same variety of disorganized tis-

Figure 7. Normal 5-day embryo implanted in uterus. Note darkly stained extra-embryonic ectoderm (upper) and embryonic ectoderm (lower) surrounded by a layer of proximal endodermal cells.
Figure 8. Early teratomas (arrows) within seminiferous tubules of a 15-day fetal mouse. Cells resemble those in inner cell mass of blastocyst.
Figure 9. Parthenote implanted in the uterus. Note ectodermal cells (ec) with proamniotic cavity surrounded by primary endoderm (en).
Figure 10. Early teratoma (arrows) with proamniotic cavity within the seminiferous tubule of a 16-day fetus. Note absence of primary endoderm.
Figure 11. Early ovarian teratoma with endoderm (en) and ectoderm (ec) surrounding proamniotic cavity.
Figure 12. Embryoid body from a transplantable testicular teratocarcinoma. Note ectoderm (ec) and embryonal carcinoma cells surrounded by endoderm (en).

sues. The early stages in development, however, appear different in some respects. During the earliest recognizable stage of testicular teratomas, there are no cells that resemble cleaving eggs. The earliest cells resemble those in the inner cell mass of normal blastocysts.

Early spontaneous and induced testicular teratomas do not form primary endoderm and trophoblast as do ovarian teratomas. On the other hand, embryoid bodies derived from

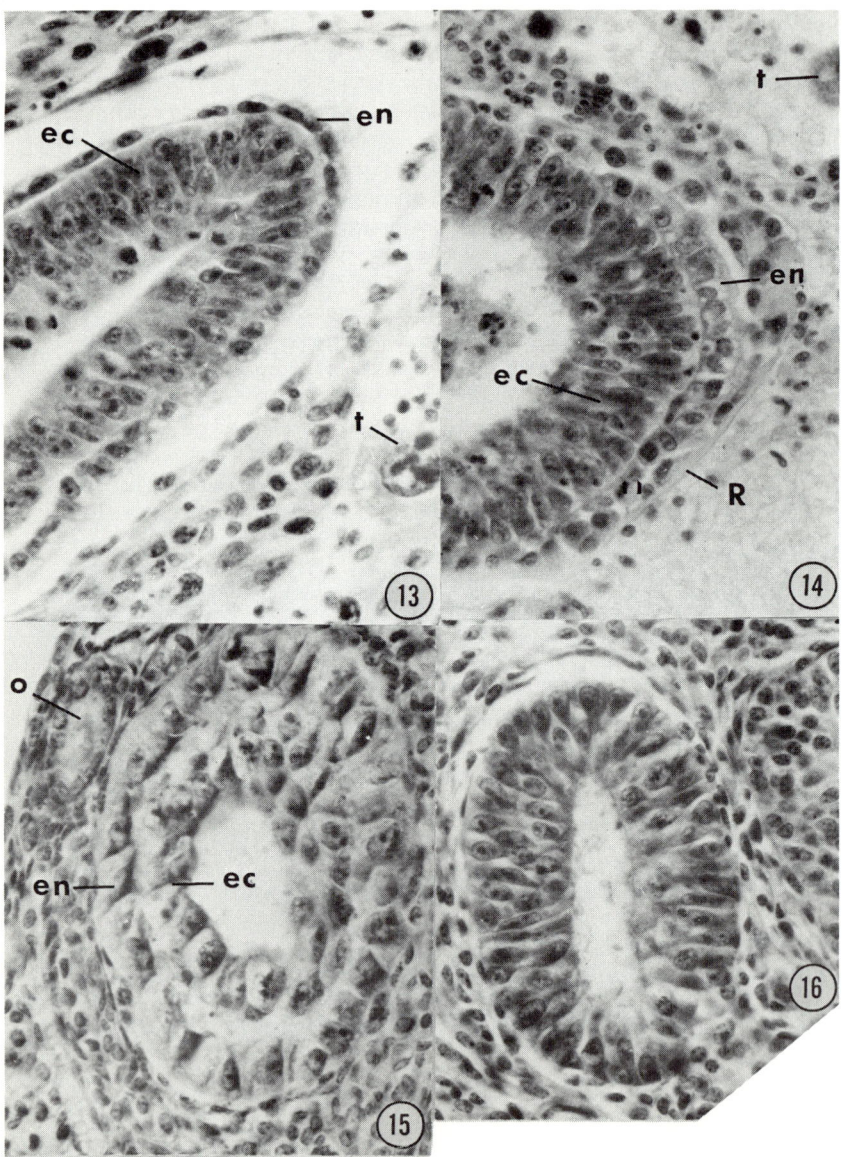

testicular teratomas do have primary endodermal cells. The primary endoderm of the rat (Levak-Svajger and Svajger, 1974) and mouse (Diwan and Stevens, 1975) appears to be transient. When dissected from egg cylinders and grafted to adult testes, it does not differentiate into other types of recognizable tissue. Early stages of testicular teratoma development contain epithelial vesicles that resemble in morphology and developmental potency the gut of 9-day embryos. Furthermore, when embryonic ectoderm is dissected from rat or mouse embryos and grafted under the kidney capsule or to the testes of adults, respectively (Levak-Svajger and Svajger, 1974; Diwan and Stevens, 1975), it forms endodermal derivatives as well as derivatives from the other two germ layers.

Ovarian teratomas originate from eggs that are activated parthenogenetically within the ovary. They develop normally for a few days as do extra-ovarian parthenotes. Normal looking blastocysts with endoderm, ectoderm, and trophoblast are formed. These ovarian blastocysts then become disorganized and undifferentiated pluripotent cells migrate into the surrounding ovarian tissue where they may continue to proliferate or they may differentiate into a wide variety of tissues.

Most parthenotes, ovarian embryos (early teratomas, Fig. 20), and testicular teratomas are disorganized by this stage of development and they lose their resemblance to normal embryos.

DISCUSSION

The comparative development of early normal and parthenogenetic mouse embryos, testicular and ovarian teratomas, and embryoid bodies from transplantable ascites teratocarcinomas has been described above. The development of strain

Figure 13. Normal 6-day embryo. Note embryonic ectoderm (ec) surrounded by primary endoderm (en). Trophoblastic giant cell (t). The clear space around the embryo is lined by Reichert's membrane.
Figure 14. Parthenote implanted in the uterus. Note embryonic ectoderm (ec), primary endoderm (en), Reichert's membrane (R), and trophoblastic giant cell (t).
Figure 15. Parthenote in ovary (early teratoma) composed of ectoderm (ec) and endoderm (en). The cells are large and probably plolyploid. Normal oocyte (o).
Figure 16. Induced testicular teratoma composed of ectoderm, but lacking primary endoderm.

LT/Sv parthenote embryos appears quite normal until about 6 days of gestation. A few survive until later stages, but most become disorganized and are aborted. They can be rescued by grafting them to the testes of adults where they grow, like teratomas, as disorganized mixtures of many kinds

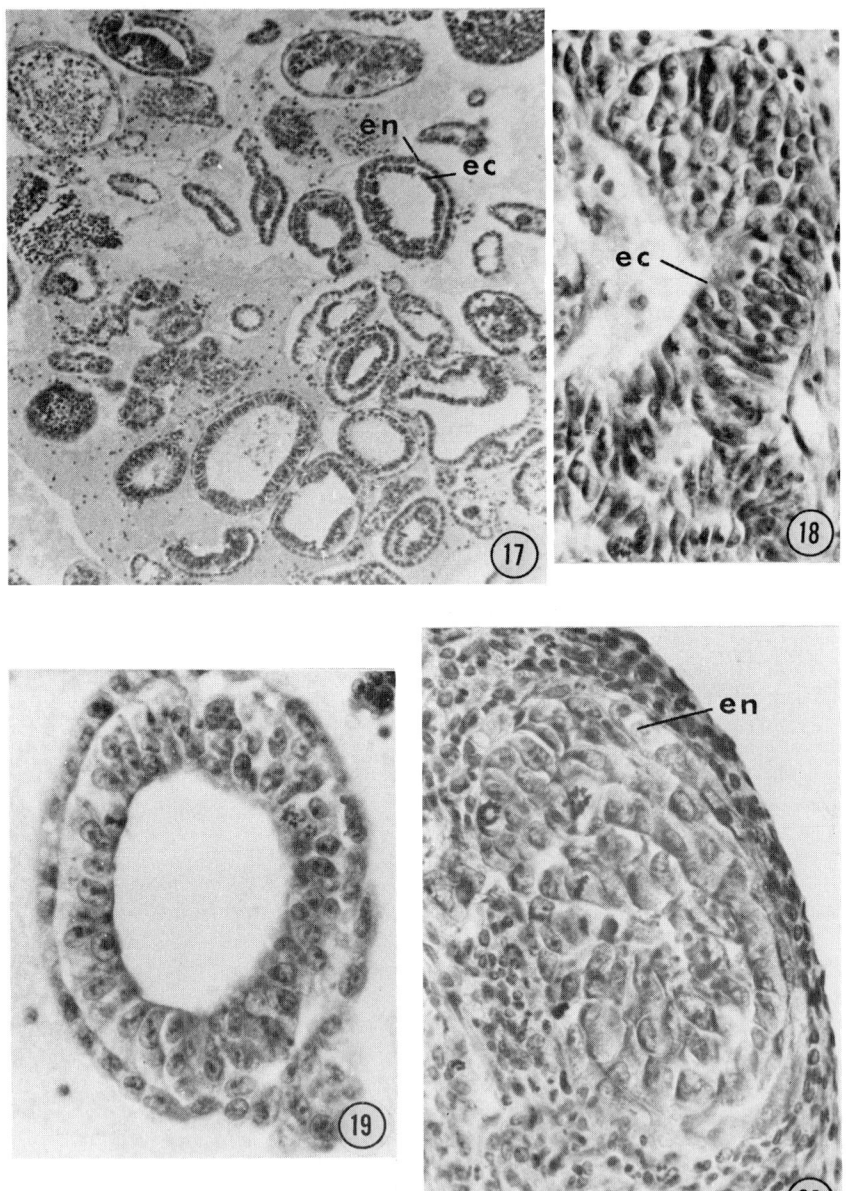

of tissues. Thus, the parthenogenetic condition in the mouse is lethal at the organismic, but not at the cellular level. Parthenogenetic blastocysts are morphologically indistinguishable from normal until they implant. Apparently, the uterus 'knows' that it is harboring something abnormal. Perhaps something could be learned about this mechanism by making chimaeras between parthenote and normal embryos or injecting parthenogenetic cells into normal blastocysts.

Testicular teratomas in mature mice resemble ovarian teratomas. They have the same variety of disorganized tissues. During early stages in development, however, they appear different in some respects. For example, in testicular teratomas there are no cells that resemble cleaving eggs as there are in ovarian teratomas. The earliest cells recognizable resemble those in the inner cell mass of normal blastocysts.

Most early testicular teratomas also differ from early ovarian teratomas in that they do not form primary endoderm or trophoblastic cells. However, a spontaneous testicular teratocarcinoma in an LG/J mouse and a transplantable tumor derived from it contained trophoblastic giant cells as well as many other cell types (Stevens and Varnum, unpublished). The primary endoderm of rat (Levak-Svajger and Svajger, 1974) and mouse (Diwan and Stevens, 1975) embryos appears to be transient. When dissected from egg cylinders and grafted to the testes of adults, it does not differentiate into mature tissues. Early testicular teratomas do form endoderm, but it resembles in morphology and developmental potency the gut of 9-day embryos (Stevens, 1959).

Even though primary endoderm is not formed by spontaneous and induced teratomas of the testes, it is present in transplantable testicular teratocarcinomas. As in normal development, the ectoderm or embryonal carcinoma of embryoid bodies is enclosed by a layer of primary endodermal cells. If the endoderm of embryoid bodies is underlaid by

Figure 17. Embryoid bodies in ascitic fluid of a mouse with a transplantable testicular teratocarcinoma. Most have an inner core of ectoderm (ec) and an outer layer of endoderm (en). One (upper left) contains fetal heart muscle. Another (upper right) contains disorganized embryonal carcinoma cells.
Figure 18. Layer of cells resembling embryonic ectoderm (ec) in an embryo-derived transplantable solid teratocarcinoma.
Figure 19. Embryoid body in ascitic fluid composed of ectoderm and endoderm.
Figure 20. Disorganized embryonal carcinoma cells partially surrounded by endoderm (en) in ovary. Probably polyploid.

mesodermal cells, yolk sac with blood islands containing immature blood cells may develop.

Testicular teratomas are similar to ovarian teratomas in that they form epithelial vesicles like the primary ectoderm surrounding the proamniotic cavity in normal 6-day

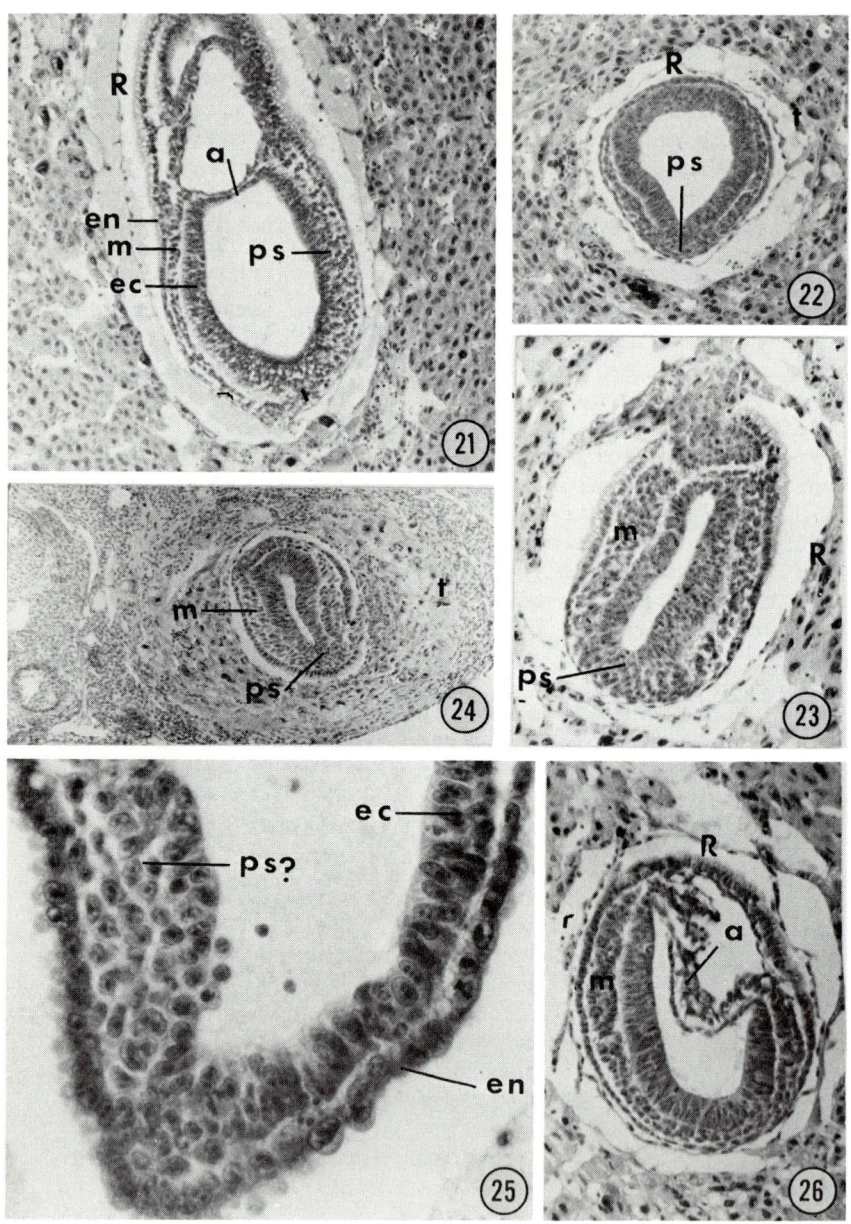

embryos. They are also similar to ovarian teratomas and parthenotes in becoming disorganized at this stage.

Specific stages in the early development of ovarian teratomas are difficult to obtain. They are microscopic and it is assumed that they develop very rapidly as do normal embryos, and the probability of finding a particular stage by histologic examination is low. Early stages in the development of testicular teratomas are relatively easy to obtain. About a third of strain 129/terSv mice have spontaneous teratomas, and they all are first recognizable histologically at 16 days of gestation (Stevens, 1973). They appear to develop at about the same rate as normal embryos.

I propose that the inner cell mass cells of normal embryos are equivalent to those in uterine parthenogenetic blastocysts, early stages of testicular and ovarian teratomas, and stem cells of progressively growing testicular and ovarian primary and transplantable teratocarcinomas including the inner cells of small simple embryoid bodies. I also propose that the primary endoderm of normal and parthenogenetic embryos is equivalent to the endoderm of embryoid bodies.

Transplantable teratocarcinomas may be derived from spontaneous primary ovarian teratomas, spontaneous and induced testicular teratomas, and teratomas derived from grafted early embryos. Teratomas from any of these sources may be considered homologous. They are all derived from undifferentiated embryonal cells. These undifferentiated cells

Figure 21. Longitudinal section of a normal 7-day embryo composed of Reichert's membrane (R), proximal endoderm (en), mesoderm (m), embryonic ectoderm (ec), amnion (a), and primitive streak (ps).
Figure 22. Cross section of a 7-day embryo with Reichert's membrane (R), proximal endoderm, mesoderm, ectoderm, and primitive streak (ps).
Figure 23. Parthenote implanted in uterus with Reichert's membrane (R), proximal endoderm, mesoderm (m), ectoderm, and primitive streak (ps). Above the embryo proper is ectoplacental cone.
Figure 24. Early ovarian teratoma (ovarian embryo) composed of trophoblast (t), Reichert's membrane, proximal endoderm, mesoderm (m), ectoderm and primitive streak (ps).
Figure 25. Embryoid body composed of endoderm (en), ectoderm (ec), and an area that resembles primitive streak (ps).
Figure 26. Uterine parthenote composed of trophoblastic cells, Reichert's membrane (R), proximal endoderm, mesoderm, ectoderm, and thick amnion (a).

originate from primordial germ cells in males, large oocytes in females, and from grafted cells of male and female embryos (Stevens, 1970b; Dunn and Stevens, 1970).

SUMMARY AND CONCLUSIONS

The comparative development of normal and parthenogenetic embryos, ovarian teratomas, spontaneous and induced testicular teratomas, and embryoid bodies is discussed. All are derived from germ cells. All early stages contain cells equivalent to the inner cell mass of normal blastocysts. All have cells equivalent to the primary ectoderm of 6-day embryos. Parthenogenetic embryos, early ovarian teratomas, and embryoid bodies derived from transplantable testicular and ovarian teratocarcinomas have transient primary endoderm, but primary and induced testicular teratomas do not. Testicular teratomas in neonatal mice have endodermal vesicles similar to the gut of 9-day normal embryos. Structures resembling the primitive streak occur in parthenogenetic embryos, early ovarian teratomas, and possibly in embryoid bodies. Transplantable teratocarcinomas may be derived from primary ovarian, primary and induced testicular teratocarcinomas and from grafted early embryos; they are all homologous.

ACKNOWLEDGEMENTS

This work was supported in part by research grant CA 02662 from the National Cancer Institute.
The Jackson Laboratory is fully accredited by the American Association for the Accreditation of Laboratory Animal Care.

REFERENCES

Diwan, S., Stevens, L. C. (1975). Development of teratomas from the ectoderm of mouse egg cylinders. Proc. Amer. Assoc. Cancer Res. 16, 98.
Dunn, G. R., Stevens, L. C. (1970). Determination of sex of teratomas derived from early mouse embryos. J. Nat. Cancer Inst. 44, 99.
Graham, C. F. (1970). Parthenogenetic mouse blastocysts. Nature (Lond.) 226, 165.
Hsu, Y.-C., Baskar, J. B. Differentiation *in vitro* of normal mouse embryos and mouse embryonal carcinoma. J. Nat. Cancer Inst. 53, 177.

Levak-Svajger, B., Svajger, A. (1974). Investigation on the origin of the definitive endoderm in the rat embryo. J. Embryol. Exp. Morph. 32, 445.
Pierce, G. B. (1961). Teratocarcinoma, a problem in developmental biology. Can. Cancer Conf. 4, 119.
Pierce, G. B., Dixon, F. J. (1959a). Testicular teratomas I. Demonstration of metamorphosis of multipotential cells. Cancer 12, 573.
Pierce, G. B., Dixon, F. J. (1959b). Testicular teratomas II. Teratocarcinoma as an ascitic tumor. Cancer 12, 584.
Pierce, G. B., Stevens, L. C., Nakane, P. K. (1967). Ultrastructural analysis of the early development of teratocarcinomas. J. Nat. Cancer Inst. 39, 755.
Stevens, L. C. (1959). Embryology of testicular teratomas in strain 129 mice. J. Nat. Cancer Inst. 23, 1249.
Stevens, L. C. (1960). Embryonic potency of embryoid bodies derived from a transplantable testicular teratoma of the mouse. Develop. Biol. 2, 285.
Stevens, L. C. (1964). Experimental production of testicular teratomas in mice. Proc. Nat. Acad. Sci. U. S. 52, 654.
Stevens, L. C. (1966). Development of resistance to teratocarcinogenesis by primordial germ cells in mice. J. Nat. Cancer Inst. 37, 859.
Stevens, L. C. (1967a). Origin of testicular teratomas from primordial germ cells in mice. J. Nat. Cancer Inst. 38, 549.
Stevens, L. C. (1967b). The biology of teratomas. Advan. Morphogenesis. 6, 1.
Stevens, L. C. (1970a). Experimental production of testicular teratomas in mice of strains 129, A/He, and their F_1 hybrids. J. Nat. Cancer Inst. 44, 923.
Stevens, L. C. (1970b). The development of transplantable teratocarcinomas from intratesticular grafts of pre- and postimplantation mouse embryos. Develop. Biol. 21, 364.
Stevens, L. C. (1973). A new hybrid subline of mice (129/terSv) with a high incidence of spontaneous congenital testicular teratomas. J. Nat. Cancer Inst. 50, 235.
Stevens, L. C. (1975). Developmental Biology of Teratomas in Mice. In "ICN-UCLA Symposium on Developmental Biology" (Ed. C. F. Fox). Academic Press, New York, in press.
Stevens, L. C. and Varnum, D. S. (1974). The development of teratomas from parthenogenetically activated ovarian mouse eggs. Develop. Biol. 37, 369.

Tarkowski, A. K. (1970). Experimental parthenogenesis in the mouse. Nature (London) 226, 162.
Theiler, K. (1972). "The House Mouse Development and Normal Stages from Fertilization to 4 Weeks of Age." Springer-Verlag, Berlin.
Whitten, W. K. (1971). Nutrient requirements for the culture of preiminplantation embryos. Advan. Biosciences 6, 129.

TERATOMAS FROM HAPLOID AND DIPLOID PARTHENOGENTIC MOUSE EMBRYOS

C. F. GRAHAM, M. W. McBURNEY AND S. A. ILES

Zoology Department, South Parks Rd.,
Oxford OX1 3PS, U. K.

INTRODUCTION

   The production of diploid teratomas and teratocarcinomas from fertilized mouse embryos is now a well established procedure which was developed by Stevens (reviewed by Stevens, 1970 and Damjanov and Solter, 1974). Preimplantation stages, egg-cylinders, and genital ridges are transferred beneath the capsule of the testis or of the kidney and develop into disorganized growths containing many differentiated cell types. If the growths contain primitive stem cells, then they are transplantable and the stem cells may retain pluripotency and a near normal karyotype both *in vivo* and *in vitro* (see other contributors to this volume).
   Our aim is to produce haploid embryos which will also form transplantable tumours which retain pluripotency and a near normal haploid karyotype. There are two methods for producing haploid mouse embryos. The most recent technique is to remove one pronucleus from the fertilized egg (Modlinski, 1975). The advantage of this method is that it allows the egg to experience fertilization and it is certain that the egg nucleus and cytoplasm are normally activated by the sperm. These haploids may contain either the female or the male pronucleus; so far they have not developed better than the haploids produced by the simpler second method. The second method is to activate the eggs with artificial stimuli; the experimenter seeks to duplicate all the effects of the sperm on the egg except those which depend on the sperm's genome. This is the method which has been used in this study.
   Several attempts have been made to produce haploid tera-

tocarcinomas from haploid parthenogenetic embryos. Graham (1970) obtained three small growths after transferring nine haploid blastocysts from F_1 C57BL/CBA mothers beneath the testis capsule, but attempts to repeat this result have always given a much lower take rate and no transplantable tumours or cell lines have been grown from such embryos. We have therefore produced growths from the parthenogenetic embryos of those mouse strains whose fertilized eggs regularly form transplantable teratocarcinomas: these are strains 129/J and C3H. This paper describes our efforts and summarises our most recent experiments (Iles et al., 1975). So far we have produced teratomas from haploid and diploid parthenogenetic embryos but these are not transplantable. The growths do, however, provide information about the capacity of the egg to support development and differentiation in the absence of the sperm.

The results which we have obtained with experimentally induced parthenogenetic embryos should be compared with spontaneous teratomas and teratocarcinomas of the ovary. The females of the LT mouse strain contain teratomas and teratocarcinomas in the ovary. When these females ovulate, spontaneous parthenogenetic development occurs and the embryos are haploid, diploid, or mosaic. It is therefore likely that the teratomas in the ovary originate from eggs which have developed parthenogenetically and which have undergone a normal or an abnormal meiosis. This view is supported by the presence in the ovary of oocytes in different stages of meiosis and of eggs in early development (Stevens and Varnum, 1974). Similarly, the teratomas of women appear to originate from parthenogenetic eggs which have undergone a meiotic division; teratomas of women who are heterozygous for particular enzyme loci may express only one of the forms of the enzyme which suggests that a reduction division has occurred (Linder and Power, 1970; Linder et al., 1975). So far the ovarian teratomas of mice and women have been reported to consist of diploid rather than haploid cells. We have, therefore, compared the development of haploid and diploid parthenogenetic mouse embryos to find out whether spontaneous haploid teratomas are likely to be found.

THE PRODUCTION OF PARTHENOGENETIC EMBRYOS

It is possible to activate unfertilized mouse eggs with a variety of techniques (reviewed by Tarkowski, 1971, 1975; Kaufman, 1975; Graham, 1974). Four of these techniques have produced parthenogenetic morulae and blastocysts (30 to 60 cell stage) which will implant. Electric shock and avertin anaesthesia can be used to activate the eggs inside the mother (Tarkowski et al., 1970; Kaufman, 1975), while heat shock and treatment with hyaluronidase are usually used *in vitro* (Komar, 1973, Graham, 1970). Electric shock produces parthenogenones which develop well; following a 30V shock to the oviduct, up to three quarters of the newly ovulated eggs are activated and over ninety percent of these will develop to the blastocyst stage (Tarkowski et al., 1970; Witkowska, 1973 a,b). Avertin anaesthesia of the mother will activate about half the eggs which are ovulated and about one quarter of these will develop to a stage at which they can induce a decidual response in the mother (Kaufman, 1975).

Heat shock, followed by treatment with cytochalasin B, produces diploid parthenogenones. Heat shock will activate up to ninety percent of eggs *in vitro* and of these, about one quarter will form morulae and blastocysts when transferred to recipient females (Balakier and Tarkowski, 1975). Hyaluronidase will activate over three quarters of eggs *in vitro* and up to three quarters of these will form blastocysts (Kaufman and Gardner, 1974; Iles et al., 1975). In these experiments we have used the hyaluronidase method of activation because it is rapid and one can observe the eggs as activation proceeds. It is also possible to control the ploidy of the parthenogenones (see below).

THE CHROMOSOME COMPLEMENT OF PARTHENOGENONES

Activated mouse eggs may be haploid, diploid, or aneuploid. The haploid embryos are produced by completion of the second meiotic division; the second polar body nucleus is either contained in a small polar body and degenerates or it is contained in a large cell which can subsequently divide and form part of the embryo. Diploid embryos are produced by the suppression of second polar body formation and the retention of both products of second meiosis in a single cell (Graham, 1971).

It is possible to control the ploidy of the parthenogenones by altering the osmolarity of the culture medium after hyaluronidase activation. In normal medium the majority of eggs develop as haploids (Graham, 1970; Graham and Deussen, 1974; Kaufman, 1973), while in low osmolar medium the majority develop as diploids (Graham, 1972; Graham and Deussen, 1974; Iles et al., 1975; Kaufman and Surani, 1974). In these experiments we therefore produced haploid and diploid parthenogenetic embryos by altering the osmolarity of

TABLE 1

ACTIVATION OF UNFERTILISED MOUSE EGGS[a]

Strain of Mother (Total Number of Eggs)	Culture Medium	% Not Activated	% Undergoing Abortive Cleavage or Lysis	Percentage of Activated Eggs			
				Haploid		Diploid	
				One Pro-nucleus 2PB	Immediate Cleavage	Two Pro-nuclei	One Pro-nucleus
C3H (1,114)	White's[b]	22.4	24.9	78.9	8.9	8.3	3.9
C3H (902)	3/5 White's	17.5	21.6	39.3	17.8	9.5	33.4
129/J (1,194)	White's	63.4	15.9	68.6	11.9	12.7	6.8
129/J (193)	3/5 White's	10.8	39.3	11.5	22.9	44.8	20.8
F₁C57BL/A2G (301)	White's	53.1	4.9	65.1	8.7	26.2	0
F₁C57BL/A2G (471)	3/5 White's	20.4	16.6	8.1	28.3	35.0	28.6
F₁C57BL/C3H (469)	3/5 White's	41.4	12.6	40.7	25.5	12.0	21.8

[a]After hyaluronidase treatment, eggs were placed in full strength White's medium or 3/5 White's medium. Two hours later they were transferred to Whitten's (1971) medium for 3 to 5 hours and then sorted out into the categories listed in the Table.

[b]see Graham, 1970.

the culture medium (Table 1). The advantage of controlling ploidy in this way is that both types of parthenogenones can be produced at the same time from a single batch of eggs.

COMPARISON OF THE PREIMPLANTATION DEVELOPMENT OF HAPLOID AND DIPLOID PARTHENOGENONES

We compared the development of haploid and diploid 129/J and C3H parthenogenones to discover whether ploidy affected development up to the blastocyst stage (approximately sixty cells). From the data in Table 2, it is clear that diploid embryos form blastocysts more often than haploids in both strains. This result is similar to that of Kaufman and Gardner (1974), who found that while fifty percent of diploid parthenogenones can induce decidualization, only thirty five percent of haploid parthenogenones are able to do so. However, the apparently superior development of the diploids does not imply that the haploid blastocysts which do develop (e.g., Fig. 1a) are unhealthy. Witkowska (1973a) counted the number of cells in haploid and diploid parthenogenones on the fourth and the fifth days of development and found that the cell numbers were similar on the fourth day but that by the fifth day the haploid morulae contained twice as many cells as the diploids and the haploid blastocysts contained one and a half times as many cells as the diploids. This high cell number in haploid embryos has been noticed previously (Beatty and Fischberg, 1951) and it suggests that the haploids which do survive until the blastocyst stage are healthy and are establishing the normal nuclear:cytoplasmic ratio. Amphibian haploid embryos also divide faster than diploids at the late blastula stage and establish a normal nuclear:cytoplasmic ratio (Graham, 1966).

There are two reasons why the diploid embryos may develop better than the haploids. It is possible that an abnormal nuclear:cytoplasmic ratio during early cleavage may disturb the development of haploids. This explanation seems unlikely to be correct because in cases in which the second polar body nucleus and the female pronucleus enter separate cells soon after activation (immediate cleavage) a normal nuclear:cytoplasmic ratio is quickly established; however, the haploid embryos formed in this way do not show superior development to those with an abnormal nuclear:cytoplasmic ratio. Another possibility is that the haploids expose recessive lethal or deleterious mutations more frequently than the diploids. This explanation is probably correct although it has not been proved. It has been shown that haploids formed from outbred

TABLE 2

DEVELOPMENT OF HAPLOID AND DIPLOID PARTHENOGENETIC
EMBRYOS AND FERTILIZED CONTROLS

Mouse Strain	Ploidy	% of Recovered Eggs Forming Morulae[a]	Blastocysts[a]	% of Transferred Embryos Forming Growths in the Kidney[b]	% of Transferred Embryos Forming Growths in the Testis[c]
C3H	Haploid	54.7 (142)[d]	3.4	37.4 (8)	3.8 (53)
	Diploid	29.5 (85)	46.6	-	3.2 (31)
	Fertilized	-	-	-	36.0 (25)
129/J	Haploid	23.2 (138)	13.8	54.8 (31)	12.4 (89)
	Diploid	34.2 (33)	31.4	-	17.4 (23)
	Fertilized	-	-	61.5 (26)	10.7 (28)
F_1C3H/ 129/J	Haploid	42.5 (73)	23.3	-	3.7 (27)

[a] The activated eggs were transferred back to the oviduct of females on the first day of pseudopregnancy. Four days later they were dissected out and cultured for a day before scoring (see Iles et al., 1975).

[b] This column summarizes experiments in which morulae and blastocysts obtained from the oviduct transfers above were transferred beneath the kidney capsule of X-irradiated syngeneic male hosts (400 to 600 rads) and inspected after 7 to 10 days. Fertilized control blastocysts were transferred to untreated kidneys of syngeneic males.

[c] Morulae and blastocysts obtained as above were transferred beneath the testis capsule of syngeneic males for 2 to 4 months.

[d] The figures in parentheses indicate the number of embryos in the samples for which the percentages were calculated.

hamster eggs develop poorly compared to those from inbred hamster eggs (Kaufman et al., 1975), but it is not clear whether the difference in development is due to genetic differences between the embryos or to cytoplasmic differences between the eggs of inbred and outbred hamster stocks. Certainly haploid embryos from eggs derived from F_1 hybrid mothers may develop to the blastocyst stage more frequently than those from inbred strain mothers (Table 2) and this effect is likely to be the consequence of differences in the cytoplasm of the activated eggs unless it is supposed that a haploid genome can show 'hybrid vigour'.

POST-IMPLANTATION DEVELOPMENT OF HAPLOID AND DIPLOID PARTHENOGENONES

Development in the Uterus

Most parthenogenetic blastocysts die soon after implantation and the most advanced embryo which has been observed to date was at the eight somite stage (Tarkowski et al., 1970; Witkowska, 1973b). This early death is unlikely to be caused by some delayed effect of the abnormal stimulus of activation; mice may form haploid embryos spontaneously (Vickers, 1969), and yet no spontaneous haploid embryos have been found late in development. It is possible that death is caused by the slow development of parthenogenetic embryos and the occurence of asynchrony between changes in the uterus and changes in the embryo. We therefore followed the procedure of Kaufman and Gardner (1974), and transferred the parthenogenones to the oviduct on the first day of pseudopregnancy, removed them on the fourth day, cultured them for one or two days, and then retransferred them to the uterus of another mother on the fourth day of pseudopregnancy. In this way the embryos were allowed to be one or two days older than the uterus close to the time of implantation. Despite this tactic, the development of the parthenogenetic embryos was poor. We transferred 81 haploid and diploid parthenogenetic morulae and blastocysts (129/J and C3H mothers) to recipients which subsequently contained at least one implantation site. Thirty-seven decidua were found but these usually contained only a few trophoblast giant cells. In three cases, delayed and abnormal seventh day embryos were dissected out; two of these were from haploid C3H parthenogenones while the other was from a diploid 129/J parthenogenone. The diploid C3H parthenogenones induced decidua more frequently than did the haploids.

Development under the Kidney Capsule

It is likely that the success of embryonic development in extra-uterine sites does not depend on the embryo undergoing developmental changes at a particular rate. We therefore transferred embryos to the kidney and found (e.g., Fig.

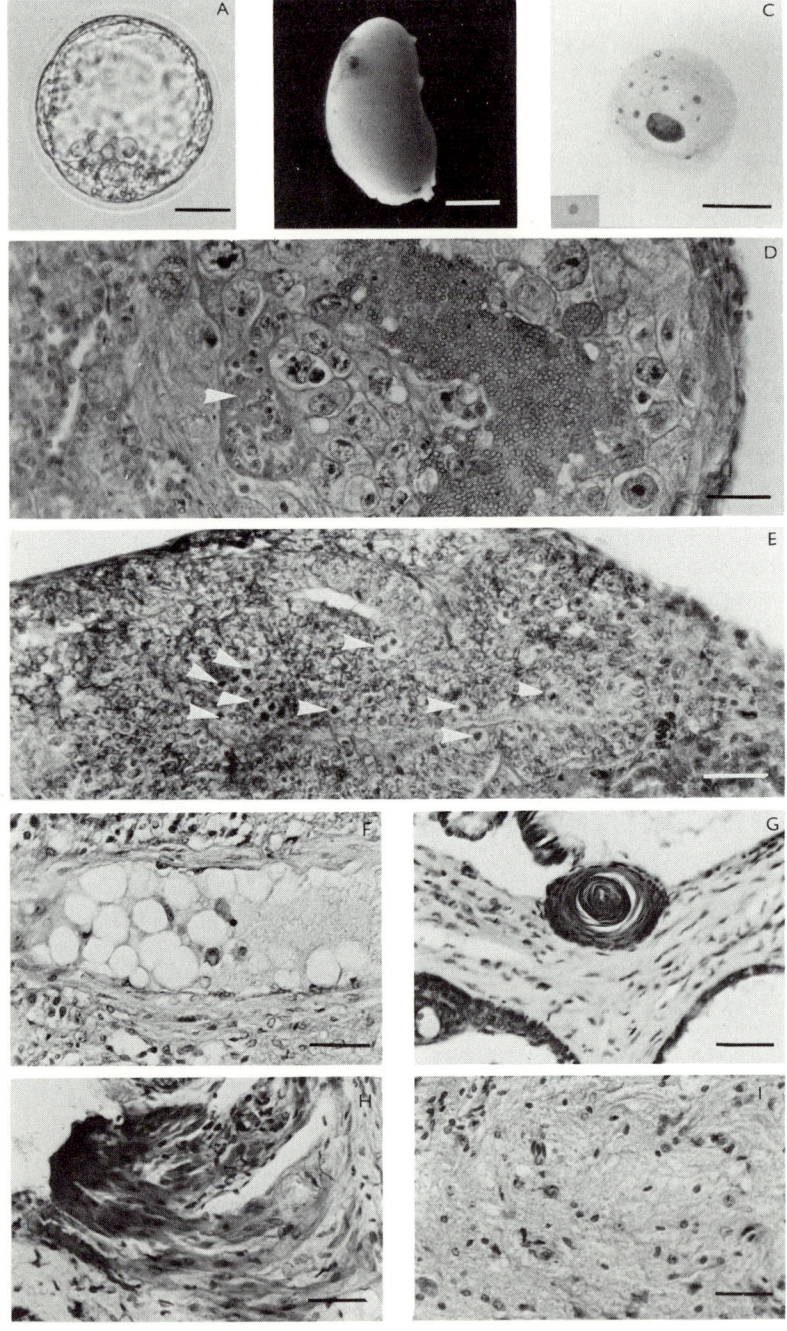

1b) that haploid blastocysts of 129/J and C3H mothers are able to grow in the kidney as are fertilized blastocysts (reviewed by Kirby, 1965). Eight growths were sectioned seven to ten days after transfer. Three contained trophoblast alone, three contained trophoblast and cells characteristic of extra-embryonic tissue, while two others contained embryonic cells as well (Fig. 1d and e). The percentage of fertilized blastocysts which form embryonic derivatives in this site is about thirty percent (e.g. Billington et al., 1968). The growths may contain many cells in mitosis (Fig. 1d) and we attempted to make chromosome preparations from twelve growths using the air drying procedure (Evans et al., 1972). Only giant and small cell nuclei in interphase were observed; the giant nuclei had diameters up to 200 μm and by reference to the relationship between nuclear diameter and DNA amount of normal mouse trophoblast (Barlow and Sherman, 1972), it is estimated that the largest contained as much as 700 times the haploid amount of DNA. There is therefore no doubt that most parthenogenetic haploid mouse embryos are

Figure 1. Growth of haploid parthenogenones in extra-uterine sites.
(a) Haploid parthenogenetic 129/J blastocyst. Scale bar = 30 um.
(b) Growth of a haploid 129/J blastocyst in the kidney (8 days after transfer). Scale bar = 0.3 cm.
(c) Giant nucleus of a trophoblast cell found in an air dried preparation of the growth shown above. The haploid cells have become polyploid. A diploid nucleus is shown at the same magnification in the inset. Scale bar = 100 um.
(d) Growth of a haploid 129/J blastocyst in the kidney (8 days after transfer). Giant trophoblast cells are obvious in this section and the arrow indicates a patch of yolk sac cells which are characterised by the presence of a thick extracellular matrix. Scale bar = 50 um.
(e) Growth of a 129/J haploid blastocyst in the kidney (8 days after transfer). Arrows indicate the numerous dividing cells which are sometimes found. Scale bar = 50 um.
(f) White adipose tissue formed by a C3H blastocyst after 2 months in the testis. Scale bar = 50 um.
(g) Keratin pearl formed by a C3H haploid blastocyst after 2 months in the testis. Scale bar = 50 um.
(h) Smooth muscle formed by a C3H haploid blastocyst after 2 months in the testis. Scale bar = 50 um.
(i) Nerve net formed by a C3H haploid blastocyst after 2 months in the testis. Scale bar = 50 um.

able to form trophoblast giant cells and that they form the same range of early embryonic tissues in similar frequency to fertilized blastocysts. Our failure to obtain chromosome preparations from these growths makes it uncertain that haploid blastocysts form embryonic egg-cylinder structures while remaining haploid.

Development in the Testis

Since even fertilized blastocysts of 129/J and C3H strains do not form transplantable teratocarcinomas in the kidney in our hands, we next transferred the parthenogenones to the testis where fertilized blastocysts are able to form transplantable teratocarcinomas. Growths are found in the testis after the transfer of haploid and diploid parthenogenetic morulae and blastocysts under the capsule; these growths are always smaller than those from fertilized controls transferred to the same site.

The growths we observed in the testis are likely to be derived from the donor embryo and not from the host tissue. This is certainly the case for C3H, F_1 C57BL/C3H, and F_1 C3H/129/J parthenogenones since the testes of these genetic stocks never exhibit spontaneous teratomas. It is possible that some of the growths in the 129/J testes were spontaneous and of host origin. We consider that the majority were derived from the donor embryos since our 129/J stocks had a low incidence of spontaneous tumours (about one in seventy), while in a particularly good transfer series we obtained eight growths from the transfer of thirty-three parthenogenetic embryos. The majority of the parthenogenetic embryos were transferred to the right testis and in the 129/J strain spontaneous tumours develop predominantly in the left testis (Stevens and Little, 1954). We therefore assume that all the growths we studied were derived from the parthenogenetic embryos.

Haploid and diploid parthenogenetic embryos formed teratomas with a similar frequency (Table 2). The C3H parthenogenetic blastocysts always appear to have a lower cell number than those of 129/J parthenogenones. The C3H parthenogenones develop poorly compared to fertilized controls from the same strain while 129/J parthenogenones form teratomas as frequently as do the fertilized blastocysts of this strain. We conclude that haploid and diploid parthenogenetic blastocysts may be able to initiate teratoma formation as well as fertilized blastocysts and this confirms the observation that the take frequency of haploid 129/J blastocysts in the kidney is similar to that of fertilized controls (Table 2).

We next studied the range of cell types which were in
these teratomas to discover an explanation for their slow
growth compared to fertilized controls. The parthenogenone
derived growths contained a similar range of differentiated
cell types to those produced by fertilized blastocysts (Fig.
1f-i; Table 3). However, the parthenogenone derived growths
lacked cells which clearly resembled yolk sac or embryonal
carcinoma and the absence of these cells probably accounts
for our failure to transplant these growths. The two growths
obtained from F_1 mothers only contained one cell type and
were very small; this is a notable result since the preimplantation development of the F_1 eggs is superior to that
of the pure strain eggs (Table 2).

THE PLOIDY OF MATURE TERATOMAS IN THE TESTIS

Our observations on the formation of trophoblast giant
cells by haploid blastocysts in the testis and the knowledge
that human amnion cells may be polyploid (Klinger and
Schwarzacher, 1958, 1960), suggest that the haploid karyotype of the blastocyst is exposed to several conditions
which promote the development of larger DNA contents during
the formation of a teratoma. We have therefore attempted to
discover the ploidy of the cells in teratomas derived from
haploid parthegenones.

Very few mitoses were observed in the teratomas formed
in the testis at the time that their hosts were killed and
we were unable to prepare reliable karyotypes directly from
the solid teratoma. We therefore established cell lines
from four of the growths and analysed their karyotypes with
G banding. In all the cell lines there was some chromosomal
rearrangement; we looked for the Y chromosome in the three
cell lines, from two haploid derived growths, which showed
the least rearrangement. The Y chromosome could not be
detected in these lines (or any of the others), and this
suggests that the cells were derived from the inevitably
female parthenogenetic embryo and not from the host testis.
In total six cell lines were established from one diploid
derived parthenogenetic growth (A), and from three haploid
derived parthenogenetic growths (B,C,D,) (Fig. 2). None of
the lines from haploid derived growths had any haploid
mitoses and line D was hypotetraploid. It was also the case
that the diploid derived growth showed two modes of mitoses
at chromosome counts of 40 and 80. We conclude that either
in the teratoma or in tissue culture there is selective
pressure for the haploid cells to increase their DNA content.

TABLE 3

DIFFERENTIATION SHOWN BY TUMOURS ARISING FROM
NORMAL AND PARTHENOGENETIC BLASTOCYSTS

Origin	Parthenogenetic						Fertilized	
Ploidy	Haploid				Diploid		Diploid	
Strain	129/J	C3H	F₁C57/C3H	F₁C3H/129/J	129/J	C3H	129/J	C3H
No. of Tumours	7	2	1	1	4	1	3	8
Embryonal Carcinoma	1?	1?	–	–	1?	1?	3	6
Neural Tissue	7	2	–	1	3	1	2	6
Pigment	3	–	–	–	1	1	1	3
Keratin. epith.	2	1	–	–	3	–	3	7
Glandular epith.	1	–	–	–	2	1	2	7
Ciliated epith.	4	2	–	–	2	1	3	6
Cartilage	2	–	–	–	1	1	1	2
Bone	3	–	–	–	2	1	3	1
Muscle	4	2	1	–	2	1	3	6
Adipose	2	1	–	–	1	–	2	3
Haemopoietic	1	–	–	–	–	–	2	–
Yolk sac carcinoma	–	–	–	–	–	–	–	1

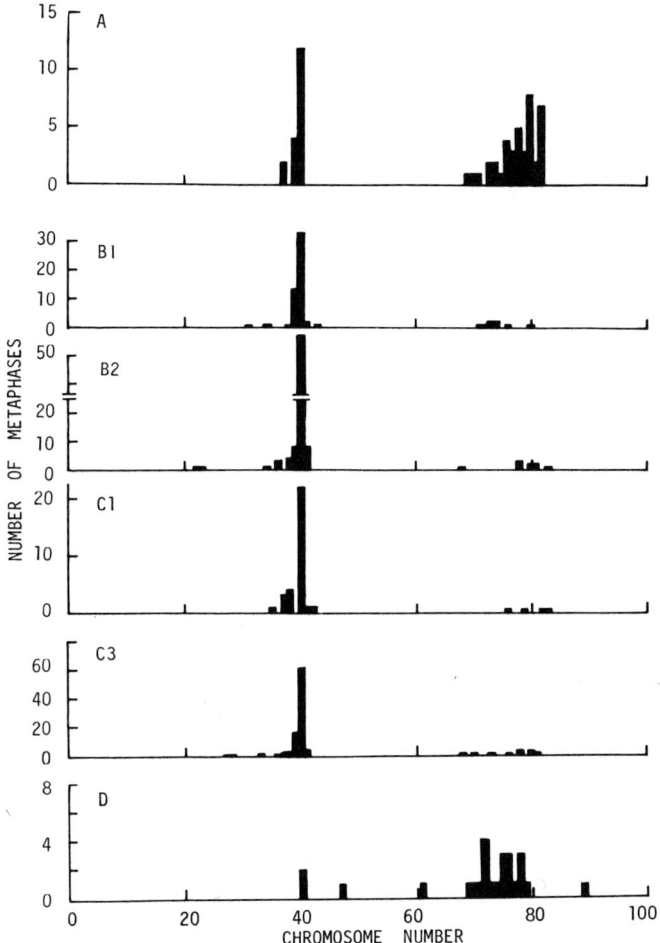

Figure 2. Frequency distribution of the chromosome numbers of cell lines derived from growths initiated by parthenogenetic embryos. Chromosomes were prepared between the second and fifth passage. A, cell line from growth of a diploid 129/J blastocyst; B1 and B2, cell lines from growth of a haploid 129/J blastocyst. These lacked a Y chromosome; C1 and C3, cell line from growth of a haploid 129/J blastocyst; D, cell line from growth of a haploid 129/J blastocyst.

The cells grown *in vitro* from the teratoma are but a small sample of the cells in the growth. We therefore measured the DNA amounts found in nuclear spreads made directly from the solid tumour (Fig. 3). It can be seen that in the haploid derived growths, more than 80% of the nuclei contained less than the normal 2C DNA amount. However, the majority of DNA values fell between 1C and 2C. Since most of the cells in the growths appeared not to be growing, it is

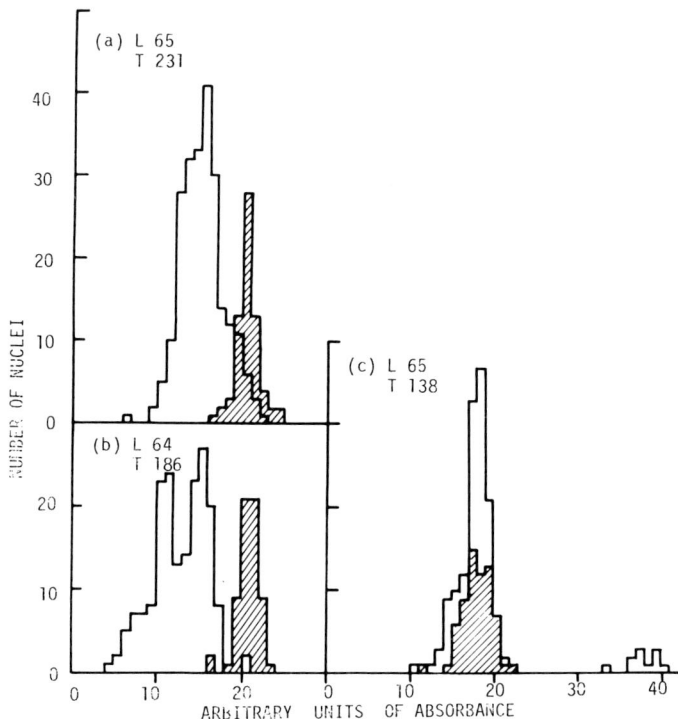

Figure 3. Microdensitometry of nuclei from growths obtained by transferring parthenogenetic blastocysts to the testis. The standard for the DNA amount of a G_1 diploid cell was provided by rat lymphocytes and their absorbance is indicated by the shaded histogram; the number measured on each slide is indicated by the figure against L. The absorbance of nuclei from the growths is indicated by the open histogram and the number measured is indicated beside T. (a) Transfer of a haploid C3H blastocyst; (b) transfer of a haploid 129/J blastocyst; (c) transfer of a diploid 129/J blastocyst.

unlikely that these cells are in the S phase and we think
that there may be a high and variable level of aneuploidy in
these cells. It is possible that chromosome non-disjunction
may occur more frequently in haploid compared to diploid
tumours.

There is no doubt that haploid and diploid parthenogenetic blastocysts can form teratomas in the testis, but we
cannot be certain that haploid blastocysts can differentiate
while remaining haploid. We have also provided evidence that
at least some of the originally haploid cells become polyploid during formation of the cell lines.

CONCLUSIONS AND DISCUSSION

The main features of teratomas derived from parthenogenetic embryos are as follows:

(a) they contain the same range of differentiated
cell types as do teratomas obtained by transferring fertilized embryos to the same site;

(b) they grow slower than fertilized controls and
they do not contain an obvious embryonal carcinoma stem cell
population. In our hands they have not proved to be transplantable although at least one spontaneous ovarian teratocarcinoma, which can be transplanted, has been found (Stevens
and Varnum, 1974);

(c) the DNA amounts in the nuclei tend to increase
above the original level. This is shown by the hyperhaploid
DNA amounts in the nuclei of the solid teratomas from haploid embryos and by the presence of diploid and hypotetraploid metaphases in the cell lines derived from these growths.

The absence of an embryonal carcinoma stem cell population may be explained in the following way. The regular
embryonic development of the parthenogenones is usually
blocked at about the seventh day of development. This may
disturb the normal production of primordial germ cells which
may be the normal progenitors of the embryonal carcinoma
cells.

Since we have demonstrated that haploid and diploid
parthenogenetic embryos can cytodifferentiate as well as
fertilized controls in these extra-uterine sites, it becomes
harder to explain the death of the parthenogenones in the
uterus. The haploids produced by Modlinski (1975) have experienced the normal activating process of fertilization and
yet they do not develop better than the haploid and diploid
parthenogenones described here. All haploid and diploid
parthenogenones and Modlinski's haploids have extensive genet-

ic homozygosity. It is, therefore, still possible to suppose that the expression of deleterious genes is the first cause of the death of the parthenogenones in the uterus. Such genes would only have to slightly slow down the development of the embryo for asynchrony to arise between its embryogenesis and the progressive changes which occur during pregnancy in the uterus. Such asynchrony is known in other circumstances to lead to the death of the embryo. This explanation of the death of the parthenogenones in the uterus also explains their superior development in extra-uterine sites where the timing of development is likely to be less important.

We have also provided evidence that haploid parthenogenetic embryos can form teratomas which contain diploid cells. The implication of this finding is that some of the spontaneous teratomas of the ovaries of mice and women may have arisen from originally haploid growths. If diploidisation occurred often in spontaneous teratomas, then this would affect attempts to use such tumours to map the human and mouse genome (e.g. Linder et al., 1975).

ACKNOWLEDGEMENTS

We would like to thank the following for advice and assistance: P. W. Barlow, S. Bradbury, S. R. Bramwell, M. D. Burtenshaw, Z. A. Deussen, E. P. Evans, M. J. Evans, S. V. Hunt, G. Martin, A. McLaren, D. Solter and A. K. Tarkowski. The Medical Research Council and Cancer Research Campaign kindly supported the work. M. W. McB. is a Canadian Centennial Postdoctoral Fellow and S.A.I. was partly supported by the Mary Goodger Scholarship.

REFERENCES

Balakier, H., Tarkowski, A. K. (1975). Diploid parthenogenetic mouse embryos produced by heat shock and cytochalasin B. J. Embryol. Exp. Morph., in press.

Barlow, P. W., Sherman, M. I. (1972). The biochemistry of differentiation of mouse trophoblast: studies on polyploidy. J. Embryol. Exp. Morph. 27, 447.

Beatty, R. A., Fischberg, M. (1951). Cell numbers in haploid, diploid, and polyploid mouse embryos. J. Exp. Biol. 28, 541.

Billington, W. D., Graham, C. F., McLaren, A. (1968). Extra-uterine development of mouse blastocysts cultured *in vitro* from early cleavage stages. J. Embryol. Exp. Morph. 20, 391.

Damjanov, I., Solter, D. (1974). Experimental teratoma. Curr. Topics Path. 59, 69.

Evans, E. P., Burtenshaw, M. D., and Ford, C. E. (1972). Chromosomes from mouse embryos and new born young; preparations from membranes and tail tips. Stain Tech. 47, 229.

Graham, C. F. (1966). The effect of cell size and DNA content on the cellular regulation of DNA synthesis in haploid and diploid embryos. Exp. Cell Res. 43, 13.

Graham, C. F. (1970). Parthenogenetic mouse blastocysts. Nature, (London) 226,165.

Graham, C. F. (1971). Experimental early parthenogenesis in mammals. Advan. Biosciences 6, 87.

Graham, C. F. (1972). Genetic manipulation of the early mouse embryo. Advan. Biosciences 8, 263.

Graham, C. F. (1974). The production of parthenogenetic mammalian embryos and their use in biological research. Biol. Rev. 49, 399.

Graham, C. F., Deussen, Z. A. (1974). *In vitro* activation of mouse eggs. J. Embryol. Exp. Morph. 31, 497.

Iles, S. A., McBurney, M. W., Bramwell, S. R., Deussen, Z. A., Graham, C. F. (1975). Development of parthenogenetic and fertilized mouse embryos in the uterus and in extra-uterine sites. J. Embryol. Exp. Morph., in press.

Kaufman, M. H. (1973). Parthenogenesis in the mouse. Nature (London) 242, 475.

Kaufman, M. H. (1975). The experimental induction of parthenogenesis in the mouse. In "The Early Development of Mammals" (Eds. M. Balls and A. E. Wild). Cambridge University Press, London, p. 25.

Kaufman, M. H., Gardner, R. L. (1974). Diploid and haploid mouse parthenogenetic development following *in vitro* activation. J. Embryol. Exp. Morph. 31, 635.

Kaufman, M. H., Surani, M. A. H. (1974). The effect of osmolarity on mouse parthenogenesis. J. Embryol. Exp. Morph. 31, 513.

Kaufman, M. H., Huberman, E., Sachs, L. (1975). Genetic control of haploid parthenogenetic development in mammalian embryos. Nature (London) 254, 694.

Klinger, H. P., Schwarzacher, H. G. (1958). Amount of sex chromatin in female tissues is correlated with degree of tissue ploidy. Nature (London) 181, 1150.

Klinger, H. P., Schwarzacher, H. G. (1960). The sex chromatin and heterochromatin bodies in human diploid and polyploid cells. J. Biophys. Biochem. Cytol. 8, 345.

Kirby, D. R. S. (1965). The role of the uterus in the early stages of mouse development. In "Pre-Implantation Stages of Pregnancy". (Eds. G.E.W. Wolstenholme and M. O'Connor). Churchill, London, p. 325.

Komar, A. (1973). Parthenogenetic development of mouse eggs activated by heat shock. J. Reprod. Fertil. 35, 433.

Linder, D., Power, J. (1970). Further evidence for post-meiotic origin of teratomas in the human female. Ann. Human Genet. 34, 21.

Linder, D., McCaw, B. K., Hecht, F. (1975). Parthenogenic origin of benign ovarian teratomas. New Eng. J. Med. 292, 63.

Modlinski, J. A. (1975). Haploid mouse embryos obtained by microsurgical removal of one pronucleus. J. Embryol. Exp. Morph., in press.

Stevens, L. C. (1970). The development of transplantable teratocarcinomas from intratesticular grafts of pre- and post-implantation mouse embryos. Develop. Biol. 21, 364.

Stevens, L. C., Little, C. C. (1954). Spontaneous testicular teratomas in an inbred strain of mouse. Proc. Nat. Acad. Sci. U.S., 40, 1080.

Stevens, L. C., Varnum, D. S. (1974). The development of teratomas from parthenogenetically activated ovarian mouse eggs. Develop. Biol. 37, 369.

Tarkowski, A. K. (1971). Recent studies on parthenogenesis in the mouse. J. Reprod. Fertil., Suppl. 14, 31.

Tarkowski, A. K. (1975). Induced parthenogenesis in the mouse. In "The Developmental Biology of Reproduction" (Eds. C. L. Markert and J. Papaconstantinou. Academic Press, New York, p. 107.

Tarkowski, A. K., Witkowska, A., Nowicka, J. (1970). Experimental parthenogenesis in the mouse. Nature (London) 226, 162.

Vickers, A. D. (1969). Delayed fertilization and chromosomal anomalies in mouse embryos. J. Reprod. Fertil. 20, 69.

Whitten, W. K. (1971). Nutrient requirements for the culture of preimplantation embryos. Advan. Biosciences 6, 129.

Witkowska, A. (1973a). Parthenogenetic development of mouse embryos *in vivo*: Preimplantation development. J. Embryol. Exp. Morph. 30, 519.

Witkowska, A. (1973b). Parthenogenetic development of mouse embryos *in vivo*: Postimplantation development. J. Embryol. Exp. Morph 30, 547.

CAN TERATOCARCINOMA CELLS COLONIZE
THE MOUSE EMBRYO?

R. L. BRINSTER

Laboratory of Reproductive Physiology
School of Veterinary Medicine
University of Pennsylvania
Philadelphia, Pennsylvania   19174

The interaction between cells in the early mammalian embryo is an important part of differentiation. This interaction has been the topic of numerous investigations which have increased our understanding of how differentiation proceeds. Studies where whole embryos are fused (Tarkowski 1961; Mintz 1962) provide one method by which the type of embryo cells interacting can be partially regulated. Another method for controlling the cells that interact was first reported by Gardner (1968, 1972). In this technique a few cells from the inner cell mass of one embryo are placed into another blastocyst. Moustafa and Brinster (1972a,b) used this technique to study the ability of older embryo cells to participate in the development of blastocysts into which they were placed. These studies indicated that cells as much as four days older than the recipient blastocyst could participate in development of the embryo.

The ability of older and asynchronous cells to participate in embryo development suggested the possibility that cells other than very early embryo cells, particularly malignant cells, might be capable of influencing the development of the embryo. To test this possibility, teratocarcinoma cells taken from ascites fluid of 129/SvSl mice, in which the tumor was propagated as an intraperitoneal tumor, were placed in mouse blastocysts.

The mouse blastocysts were obtained by superovulating random-bred Swiss albino females with gonadotrophins (Brinster 1963, 1972). On the fourth day after mating, blastocysts were collected by flushing the uterus with culture medium. Eagle's basal medium supplemented with 5×10^{-4}M pyruvate and 10% fetal calf serum (EBMS) was used. The blastocysts from several females were pooled, washed three times, and stored in medium at 37° under an atmosphere of 5% CO_2 in air (Brinster, 1972). These blastocysts were the recipients for the transferred cells.

Cells from mouse teratocarcinoma (OTT 6050) were obtained from 129/SvSl mice carrying the tumor in ascitic form (Stevens and Little, 1954; Stevens, 1970). In the ascitic fluid are found cellular aggregates which contain embryonal carcinoma cells enveloped in a layer of endoderm (Stevens, 1970). These embryoid bodies were placed in phosphate buffered saline containing 0.25% trypsin and separated into individual cells by pipetting them back and forth in a narrow pipette. The cells were placed in Eagle's basal medium with 10% fetal calf serum.

The apparatus used to make the cell transfers consists of right and left handed Leitz micromanipulators and a Leitz Laborlux II microscope. The blastocysts were placed in the well of a cavity slide containing 100 µl of the culture medium (EBMS with the $NaHCO_3$ replaced by HEPES) and covered by liquid silicone. The left hand micromanipulator holds a blunt suction pipette (the inside diameter is 5 µm and the outside diameter is 100 µm) and the right hand micromanipulator holds the injection pipette. The diameter of the injection pipette is varied to just accommodate the type of cell to be injected. The pipette should have a thin wall and a tapered tip. The blastocysts and cells are placed in the well of the cavity slide, and the blastocyst picked up on the end of the suction pipette. The cells are picked up with the injection pipette (previously filled with oil) and the pipette inserted through the zona pellucida and blastocoele wall into the cavity. Two to four cells are placed near or against the inner cell mass. The pipette is slowly withdrawn to prevent the escape of the cells. The injected or recipient blastocyst is then removed from the slide and placed in a watch glass containing culture medium until transfer to a foster mother.

The foster mothers are obtained by injecting random-bred Swiss albino females with 2 units of pregnant mare serum followed in 48 hours with an injection of 2 units of human chorionic gonadotrophin. The females are placed with vasectomized males who have been proven sterile. The injections for the foster mothers are begun 24 hours after those for the females producing the recipient blastocysts. The foster mothers are anesthetized with pentabarbitol and the hair clipped from a 3 centimeter square area over the area of the ovary. A one centimeter incision is made in the skin and abdominal wall over the ovary and the ovarian fat pad, ovary and uterus withdrawn. The recipient blastocysts are picked up in a small volume of culture medium in a finely drawn pipette. The pipette is inserted through the uterine wall and the blastocysts and medium are gently forced into the uterine lumen of each foster mother. The incision is closed and the female placed in a cage alone until the young are born.

The results of the cell transfers were determined in three ways. First, the animals born to the foster mothers were observed for pigmentation of eye or coat color. The recipient blastocysts were always albino and the cells (129/SvSl) came from agouti animals. Second, the offspring resulting from blastocysts into which cells were transferred were placed with albino animals of opposite sex for mating. The young from these matings were examined to determine if any germ cells contained agouti genotype from the transferred cells. Third, skin grafts were placed according to the technique described by Billingham (1961) on the young resulting from the blastocysts which received teratocarcinoma cells. The survival time of these skin grafts was compared to the survival time of similar grafts placed on control animals from the same strain as those supplying the recipient blastocysts (ICR Swiss albino males or females). Median survival times, 95% confidence limits, and statistical significance were determined according to the methods described by Litchfield (1949; Litchfield and Wilcoxon, 1949).

From blastocysts that received teratocarcinoma cells, a total of 60 offspring were reared to adults (Brinster, 1974). Thirty were males and 30 were females. One of the males had several small patches or stripes of agouti hair. Fig. 1 shows the extent of these agouti areas. The areas resembled thin stripes of hair running from the dorsal midline laterally on the right thorax and flank. They did not extend across the dorsal midline or on to the belly hair. The pattern was similar to that seen in some animals resulting from whole embryo fusion (Mintz, 1970). There was also a small area of agouti hair immediately anterior to the base of the left ear. This male was mated to numerous females but no offspring resulted. Skin from a 129/SvSl male was grafted on this male. The first graft was torn off by the animal shortly after the protective case was removed. The second graft remained for 26 days before it was off.

Skin grafts were made on all 60 of the animals resulting from the blastocysts that received 129/SvSl cells, and there was a significant increase in survival time of these grafts when compared to those made on control animals. These results are shown in Table 1. The graft median survival time for experimental males and females was 11.0 and 8.9 days respectively. Both these values are significantly greater than the corresponding controls. In addition, the median graft survival time on the males was 2.1 days longer than on the females. This difference was significant, but the difference in median graft survival time for the male and female controls was not significant.

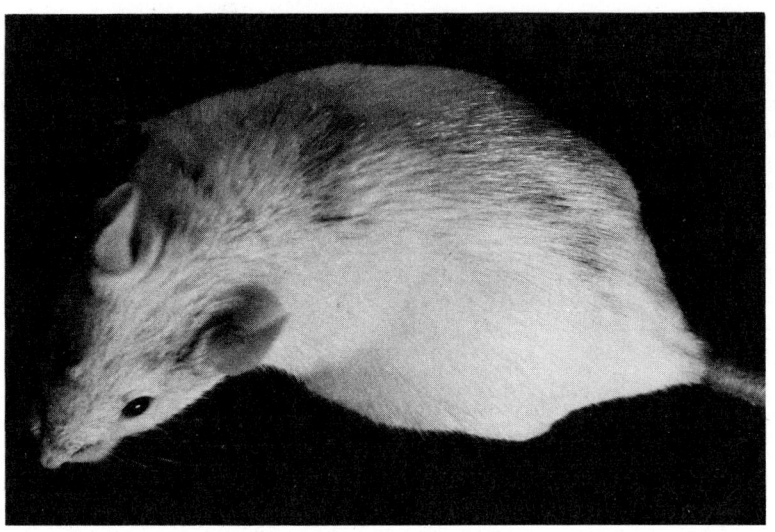

Figure 1. A Random-bred Swiss albino mouse with several stripes of agouti hair extending from the midline over the right flank and thorax. An additional group of agouti hairs is present at the anterior base of the left ear. The animal developed from a blastocyst of Swiss albino genotype that received 129/SvSl (agouti) teratocarcinoma cells (Brinster, 1974).

The distribution of graft survival time for experimental animals was quite skewed (see Fig. 2). Thirty-six rejections occurred between 7 and 10 days; twenty-one between 11 and 20 days; and 3 after 20 days.

All of the animals resulting from the teratocarcinoma cell transfers were placed with albino mice to determine if there were any germ cells with agouti genotype. No agouti young resulted from these matings.

The results of these experiments in which 129/SvSl teratocarcinoma cells were transferred into blastocysts provide

TABLE 1

EFFECT OF CELLS TRANSFERRED INTO MOUSE BLASTOCYSTS ON THE IMMUNE RESPONSE OF OFFSPRING RESULTING FROM THE RECIPIENT BLASTOCYSTS

Cell Type Transferred	Skin Graft Median Survival Time	
	Males	Females
Teratocarcinoma Cells	11.0 [9.6-12.6] (30)	8.9 [8.1-9.7] (30)
None (Control Animals)	6.8 [6.0- 7.7] (30)	7.8 [7.3-8.3] (30)

Median survival time is given in days and is followed by the 95% confidence limits enclosed in brackets. The number of animals in each group is shown in parentheses. The significance of the difference between median survival times was calculated by the method of Litchfield (1949; Litchfield and Wilcoxon, 1949). The increase in median survival time was significant at the P<0.05 level for the groups that had received cells. The difference between males and females was significant only for the animals that received teratocarcinoma cells but not for the controls.

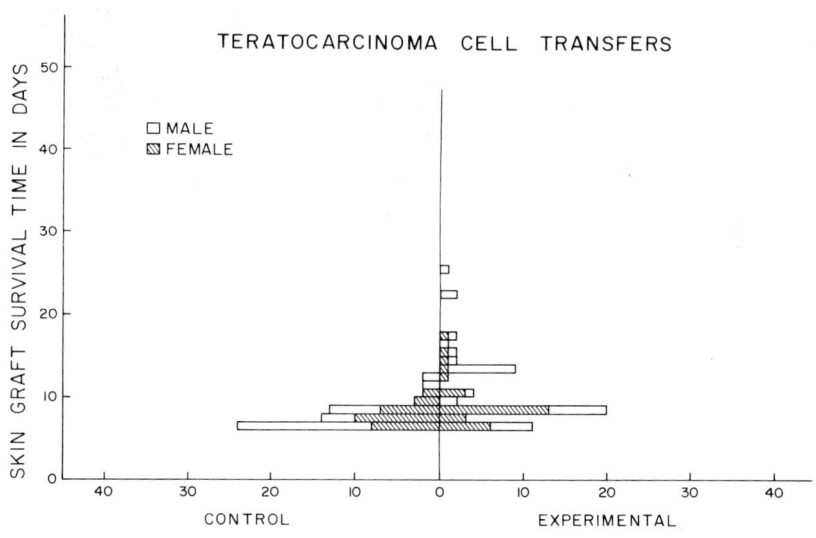

Figure 2. *The distribution of skin graft survival times on male and female mice that resulted from blastocysts receiving teratocarcinoma cells (Experimental), and survival times on comparable animals that were never exposed to teratocarcinoma cells (Control).*

evidence that this type of malignant cell is capable of participating in development of the embryo. In addition, the experiments indicate that the embryo environment can bring under control the autonomous proliferation of teratocarcinoma cells. The interaction of the teratocarcinoma cells with embryo cells must be quite different than with adult cells, since in the adult the cells generally result in death of the animal in approximately three weeks.

The appearance of the agouti hairs in one of the offspring resulting from the 129/SvSl cell transfers provides very strong evidence for the participation of the transferred cells in embryo development. Although the number of hairs is small, they are unequivocably present and could not arise from the albino recipient blastocyst genotype. The inability of the animal to maintain a 129/SvSl skin graft indefinitely could have been due to the type of cell or small number of cells present. These may not have created sufficient tolerance to maintain a skin graft. A number of situations have been described in which chimerism exists without skin graft tolerance (Billingham and Silvers, 1971).

The skin graft data from the animals that received the teratocarcinoma cell transfers provide additional and convincing evidence for the participation in embryo development and for the presence in the adult of 129/SvSl cells. The reason for the 2.1 day difference in graft survival time between the males and females is not readily apparent. A possible explanation may be related to the sex genotype of the transferred cells. The teratocarcinoma cells used in these studies carried the Y chromosome (Stevens, 1970). Perhaps it is more difficult for a cell with a Y chromosome to establish a colony in a female embryo because of antigens associated with Y chromosome expression. This could explain why the cloning of the females was less efficient than the cloning of the males by the male genotype teratocarcinoma cells. A similar effect was not seen when CBA bone marrow cells were transferred into blastocysts (Brinster, 1974), possibly because male and female cells were injected into male and female blastocysts randomly. This would tend to obscure the effect of the antigens of the Y chromosome. However, it has been demonstrated that the Y antigen has only a weak effect (Gasser and Silvers, 1972), and the difference in median survival time between the male and female groups may result from other causes.

ACKNOWLEDGEMENTS

The work described was supported by Public Health Service Research Grant No. CA 14676 from the National Cancer Institute and HD 08539 from the National Institute of Child Health and Human Development. I thank Dr. Roy Stevens for the 129/SvSl teratocarcinoma cell line and Dr. Willys Silvers for instruction in skin grafting of mice. In addition I thank Drs. Alan Beer, Lionel Manson, Joy Palm, Willys Silvers, Roy Stevens and Wes Whitten for helpful suggestions.

REFERENCES

Billingham, R. E. (1961). Free skin grafting in mammals. In "Transplantation of Tissues and Cells" (Eds. R. E. Billingham and W. K. Silvers). The Wistar Institute Press, Philadelphia, p. 1.
Billingham, R. E., Silvers, W. K. (1971). Immunological tolerance. In "The Immunobiology of Transplantation" (Eds. A. Osler and L. Weiss). Prentice-Hall, Englewood Cliffs, New Jersey, p. 132.
Brinster, R. L. (1963). A method for *in vitro* cultivation of mouse ova from two-cell to blastocyst. Exp. Cell Res. 32, 205.
Brinster, R. L. (1972). Cultivation of the mammalian egg. In "Growth, Nutrition and Metabolism of Cells in Culture" (Eds. G. Rothblat and V. Cristofalo). Academic Press, New York, p. 251.
Brinster, R. L. (1974). The effect of cells transferred into the mouse blastocyst on subsequent development. J. Exp. Med. 140, 1049.
Gardner, R. L. (1968). Mouse chimaeras obtained by the injection of cells into the blastocyst. Nature 220, 596.
Gardner, R. L. (1972). An investigation of inner cell mass and trophoblast tissues following their isolation from the mouse blastocyst. J. Embryol. Exp. Morph. 28, 279.
Gasser, D. L., Silvers, W. K. (1972). Genetics and immunology of sex-linked antigens. Advan. Immunol. 15, 215.
Litchfield, J. T., Jr. (1949). A method for rapid graphic solution of time-per cent effect curves. J. Pharmacol. Exp. Therapy, 97, 399.
Litchfield, J. T., Jr., Wilcoxon, F. (1949). A simplified method of evaluating dose-effect experiments. J. Pharmacol. Exp. Therapy 96, 99.
Mintz, B. (1962). Formation of genotypically mosaic mouse embryos. Amer. Zool. 2, 432.

Mintz, B. (1970). Gene expression in allophenic mice. Symp. Intern. Soc. Cell Biol. 9, 15.

Moustafa, L. A., Brinster, R. L. (1972a). The fate of transplanted cells in mouse blastocysts. J. Exp. Zool. 181, 181.

Moustafa, L. A., Brinster, R. L. (1972b). Induced chimaerism by transplanting embryonic cells into mouse blastocysts. J. Exp. Zool. 181, 193.

Stevens, L. C., Jr., Little, C. C. (1954). Spontaneous testicular teratomas in an inbred strain of mice. Proc. Nat. Acad. Sci. U.S. 40, 1080.

Stevens, L. C., Jr. (1970). The development of transplantable teratocarcinomas from intratesticular grafts of pre- and post-implantation mouse embryos. Develop. Biol. 21, 364.

Tarkowski, A. K. (1961). Mouse chimaeras developed from fused eggs. Nature 190, 857.

DEVELOPMENTAL AND EXPERIMENTAL POTENTIALITIES OF MOUSE TERATOCARCINOMA CELLS FROM EMBRYOID BODY CORES

BEATRICE MINTZ, KARL ILLMENSEE, AND
JOHN D. GEARHART

Institute for Cancer Research
Fox Chase
Philadelphia, Pennsylvania   19111

Mouse teratocarcinoma, or embryonal carcinoma, cells, with their multitude of developmental capacities (Stevens, 1967; Pierce, 1967), present many unique prospects for the experimental study of mammalian genetic regulatory mechanisms governing differentiation and carcinogenesis. Several years ago, when work along such lines was being contemplated in this laboratory, the first author concluded that an *in vivo* source of the carcinoma cells would be much more promising than an *in vitro* source for the particular long-range aims of our program.

These aims relied heavily on the projected use of mutagenized teratocarcinoma cells. Just as mutants once provided the chief tool for probing regulatory systems in microbial organisms, they could provide an equally powerful one with which to reveal mammalian regulatory systems. The plan was, first, to identify a source of embryonal carcinoma cells whose developmental potentialities could be experimentally shown to comprise virtually all paths of tissue specialization. These cells would then be the 'theme' on which mutational 'variations' might be produced and selected for in short-term cultures. Some of the most interesting mutations would be expected to be those affecting differentiation. But the maximal usefulness of such mutations could only be realized if the mutant cells could be tested in the context of development of the entire organism. While short-term cell cultures of 'parent' and of mutant clones could furnish valuable biochemical and other sorts of information, the most far-reaching 'differentiation phenotypes' would be well beyond the scope of realization in culture or even in the chaotically arranged solid tumors.

It should also be pointed out that it is very difficult to prove that a 'variant' clone of somatic cells in culture has actually been changed genetically, as opposed to occurrence of variation for other reasons. Orderly recombination and segregation of mammalian genes takes place only in meiosis in the germ line; it is this recombination and transmission that will demonstrate whether a phenotype is due to a definable genetic unit, which can then be mapped.

The decision to rely on an *in vivo* source of embryonal carcinoma cells was based on the presumption that chromosomal alterations and indefinable genetic changes would be much more likely to accumulate during long-term cultivation of cells *in vitro* than during their long-term transplantation *in vivo*; such changes would make it far more difficult to realize our specific long-range goals, as stated above. Teratocarcinoma cultured cell lines conveniently lend themselves to the pursuit of many important questions, some of which might not otherwise be amenable to investigation. However, most of these lines do not in fact have a normal chromosome number (e.g., Kahan and Ephrussi, 1970; Evans, 1972; Jakob et al., 1973; Lehman et al., 1974), whereas the tumors from which they originated did have a diploid karyotype (Dunn and Stevens, 1970). The direct cellular transplant descendants of the latter were therefore thought to have possibly retained relative normalcy of chromosomal constitution and, thus, to be appropriate for experiments ultimately requiring 'cycling' of totipotent cells into developing embryos.

TERATOCARCINOMA CELLS FROM EMBRYOID BODY 'CORES'

Of the two presently available *in vivo* sources of teratocarcinoma cells, the solid tumors growing attached to a substratum (e.g., in a subcutaneous site) were deemed unsuitable because the malignant multipotential stem cells are erratically intermingled with their benign differentiated cellular progeny. The other *in vivo* source--the embryoid bodies in ascites suspensions propagated by intraperitoneal transfers--appeared quite promising because the location of multipotential cells in them is known. Small-size embryoid bodies (Fig. 1) comprise a 'core' of embryonal carcinoma cells surrounded by an epithelial 'rind' of yolk-sac-like cells; older, larger embryoid bodies often become cystic and may show early internal differentiation (Stevens, 1967; Pierce, 1967). The multipotentiality of single cells from completely dissociated embryoid bodies has been demonstrated by formation of a variety of tissues in tumors produced when such cells were inoculated into animals (Kleinsmith and Pierce, 1964).

We are now attempting to obtain a pure *in vivo* ascites line of transplantable cores (and possibly eventually of totipotent unaggregated core cells) in large yield. (While cores have often been observed to form in cultured teratocarcinoma lines, the term as used here will refer only to cores always grown *in vivo*.) These efforts, while still in a preliminary stage with limited yield, appear promising. In those studies to be presented here in which core cells were involved, relatively small numbers of these cells were used and they were freshly obtained directly from small-size embryoid bodies subjected to brief proteolytic treatment (in 0.25% trypsin and 1.25% pancreatin) and then 'peeled' with tungsten needles. The embryoid bodies are from the OTT 6050

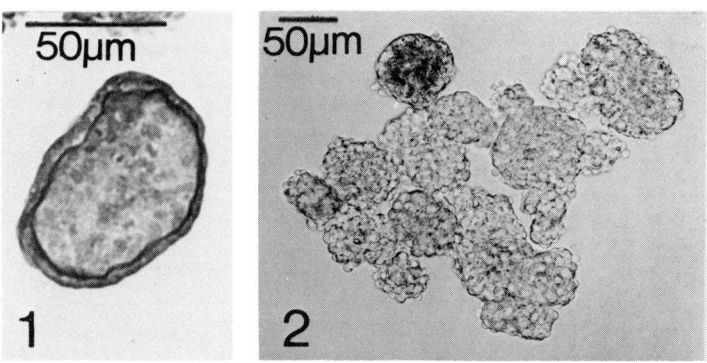

Figure 1. Small-size OTT 6050 embryoid body, sectioned and stained with the periodic acid-Schiff reagent. The core of embryonal carcinoma cells is surrounded by a yolk sac epithelial rind; basement membrane secreted by the latter separates the two parts.
Figure 2. Living cores (clear) of embryonal carcinoma cells, and an intact embryoid body (dark), for comparison.

ascites teratoma (Stevens, 1970), originally received from Dr. L. C. Stevens and maintained by intraperitoneal transfers every 2-3 weeks in syngeneic males of the 129/Sv Sl^J C P inbred strain (to be referred to as 129). Peeled cores (Fig. 2) isolated in this manner and explanted (in Dulbecco's modification of Eagle's medium plus 10% heat-inactivated fetal calf serum) showed good viability (Fig. 8) and proliferated cells of various morphological types, with myogenic cells becoming particularly conspicuous and forming myotubes (Fig. 9), as already reported for cultures of intact embryoid bodies (Gearhart and Mintz, 1974, 1975).

Chromosome counts of metaphase cells in colchicine-treated, air-dried preparations of cores were obtained with the assistance of Mrs. Claire Cronmiller in this laboratory. Results confirmed the expectation that the multipotential core cells might have a euploid chromosome complement despite their 8-year-long transplant history. As seen in Fig. 3, there is a high modal frequency of cells with 40 chromosomes, and only

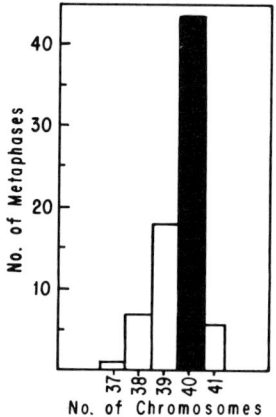

Figure 3. The modal normal chromosome number of 40 in metaphases of embryonal carcinoma cells of cores from OTT 6050 embryoid bodies grown only in vivo *as an ascites for 8 years.*

a very narrow range of deviation in chromosome numbers. Thus, the low incidence of euploidy and the great variability in chromosome number previously observed in direct preparations of whole embryoid bodies (Jakob et al., 1973; Lehman et al., 1974; Nicolas et al., 1975) may be due to their rinds of yolk sac cells, which are not developmentally multipotential, and/or to inclusion of older cystic embryoid bodies in the preparations.

TOTIPOTENCY AND NORMALIZED GENE EXPRESSION OF MALIGNANT TERATOCARCINOMA CELLS FROM EMBRYOID BODY CORES

While the multipotentiality of teratocarcinoma stem cells is evident from their differentiation into a variety of tissues in tumors, there has heretofore been no indication that the cells are capable of giving rise to all cell types. The absence of certain tissues in the many solid tumors examined (Stevens and Hummel, 1957) is consistent with the possibility that the stem cells are not comparable to totipotent early embryo cells but rather to some later stage of embryo cell in which the range of developmental potentialities has become restricted. Alternatively, the failure to form certain tissues in the tumors might result from the chaotic distribution of particular cells whose association and interaction might be a prerequisite.

The only conclusive way to ascertain the full developmental capacities of embryonal carcinoma cells would be to test whether, in a more orderly environment appropriate to embryogenesis, they would be capable of giving rise to all tissues. Such an experiment would be analogous to producing an allophenic mouse (Mintz, 1971b), except that the latter is formed by artificially associating totipotent cells from two genetically different normal embryos, whereas the present experiment involves an association of normal embryo cells and putatively totipotent embryoid body core cells. If this experiment were to succeed, it would provide a valid base for proceeding with similar tests of mutagenized core cells. Full gene expression in the tumor-derived cells would also demonstrate that all the genetic information required to make a normal mouse was present in the carcinoma cells, hence that their initial conversion to malignancy had not involved mutational change.

Preliminary trials by the aggregation method (Mintz, 1971b) showed that core cells did not adhere well to blastomeres of mid-cleavage or morula stages and thus differed from the latter in some relevant surface properties. Therefore, the method of injection into the blastocyst cavity (Lin, 1966; Gardner, 1968) was used in order to entrap the core cells in the embryo. Blastocysts with many genetic markers distinguishing them from the strain of origin of the core cells were flushed out of the uteri on day 3 of pregnancy (counting the plug date as day 0), and were placed in a microdrop of medium under oil on a Zeiss inverted microscope. Each blastocyst was then held on the end of a blunt-tipped siliconed pipette in a Leitz micromanipulator and injected in the blastocoel cavity near the inner cell mass with 5 core cells from a pointed pipette. After a few hours' incubation, the injected blastocysts were surgically transferred to uteri of pseudopregnant females (usually of the ICR random-bred albino strain) mated to sterile vasectomized males a day later than the matings of the blastocyst donors.

Of 280 blastocysts injected, 45 living embryos or fetuses were obtained from prenatal autopsies and 48 mice were born, all live. No abnormalities or tumors were evident. Of 14 individuals thus far analyzed for presence of 129-strain cells, 3 postnatal animals have proved to be cellular genetic mosaics with appreciable tissue contributions from the teratocarcinoma cell strain. These three will be briefly described here. More detailed descriptions of them will be presented elsewhere (Mintz and Illmensee, 1975) and analyses of the remainder of the series will be reported subsequently.

The first experimental animal (*mosaic mouse #1*), shown in Fig. 4, is a male, still alive and healthy at 9 weeks of age. (Mice inoculated with OTT 6050 embryonal carcinoma cells ordinarily die after 3-4 weeks). He is from a C57BL/6-*b/b* blastocyst, coisogenic with C57BL/6, with *b* (*brown*) genes substituted for *B* (*black*) as a result of a

Figure 4. A genetically mosaic male mouse (mouse #1) obtained from a C57BL/6-b/b *(non-agouti brown) blastocyst injected with embryonal carcinoma cells from cores of OTT 6050 embryoid bodies of strain 129/Sv* Sl^J *C P (agouti black). The coat is largely agouti black and is almost entirely derived from the tumor-cell strain. Fine transverse stripes of agouti and non-agouti hair follicle clones are present on each flank and on the crown of the head. Brown light patches may be seen over each haunch and on the head. The overall 'diluted' appearance of the coat is due to the steel gene (Sl^J/+), whose presence in the carcinoma cells was previously unknown and was revealed by the coat phenotype in this animal.*

mutation at that locus. The embryo strain is therefore $a/a;b/b$ (*non-agouti* and *brown*), the 129 tumor strain $Aw/Aw;B/B$ (*white-bellied agouti* and *black*). The experimental animal is extremely striking because he greatly resembles 129-strain control mice. His coat is largely *agouti* and *black*, and he also has the cream-colored belly characteristic of A^w/A^w. However, he is in fact partially striped, as with allophenic mice formed from normal blastomeres of such paired color strains and fortuitously having a predominance of the *agouti black* cellular component (Mintz, 1967, 1970a, 1971a, 1974). The transverse fine-striped areas on both flanks and on the crown of the head represent the dermal component of clones of *agouti* or *non-agouti* hair follicles, each presumably descended from a somite dermatome cell whose mitotic progeny spread laterally. Each *brown* patch, over each haunch and on each half of the head, represents a surviving part of a melanoblast clone that had migrated laterally from a neural-crest-derived cell on that side of the midline. The tail is also striped with *black* and *brown* melanoblast bands.

Genes at the *agouti* locus are expressed in hair follicle cells, not in melanoblasts, but the phenotype of the hair follicle environment then determines whether the melanoblasts will produce eumelanin or a yellow modification, phaeomelanin, as in the subterminal yellow bands of *agouti* hairs (Silvers and Russell, 1955). Even in the *brown* regions of mouse #1, the hairs are chiefly *agouti* and the resultant color in those areas therefore differs from the true brown of C57BL/6-*b/b* controls.

As will be discussed in more detail in the following section, the coat of this animal also showed the phenotype produced by the *steel* (Sl^J) gene.

The circulating red blood cells and also the white blood cells of mouse #1 were almost entirely of the 129-strain type (Fig. 5), judging from their proportions of strain-specific allelic variants of glucosephosphate isomerase (*Gpi-1* locus), as analyzed with a starch gel electrophoresis method (Gearhart and Mintz, 1972). As would be expected from their genotypes, the erythrocytes produced chiefly the type of hemoglobin appropriate to the 129 strain, i.e., the *diffuse* (Hbb^d allele) (Ranney et al., 1960) rather than the C57BL/6 or *single* (Hbb^s) type.

Ouchterlony double diffusion analyses for allotypes of 7S serum antibodies of the G1 and G2a immunoglobulin classes (*Ig-1* and *Ig-4* loci) were kindly conducted by Dr. Melvin J. Bosma of this Institute with reagents already described (Bosma and Bosma, 1974). The experimental animal's allotypes were entirely of the 129-strain variety.

The major urinary protein complex (*Mup-1* locus) normally excreted by mice is manufactured in the liver (Finlayson et al., 1965). The proportion of two allelic strains of the protein in the urine has been found to reflect the proportions of the corresponding cell strains of hepatocytes in a given allophenic mouse (Condamine et al., 1971). Starch gel electrophoretic separation (Condamine et al., 1971) of the urinary protein of mouse #1 (Fig. 6) thus documents the presence of substantial numbers of 129-strain hepatocytes in his liver.

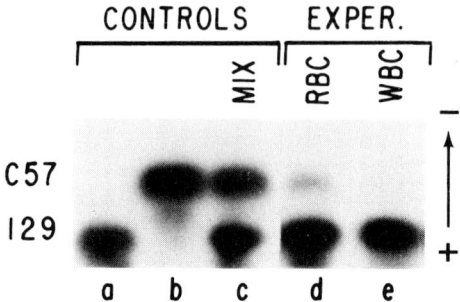

Figure 5. *Strain-specific allelic variants of glucosephosphate isomerase in starch gel electrophoresis of blood cell lysates from 2 pure-strain controls (slots a, b), a control mixture (c), and the red (d) and white (e) blood cells of the experimental mouse shown in Fig. 4.*

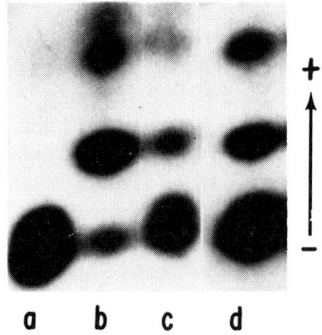

Figure 6. *Starch gel electrophoretic patterns of allelic variants of the major urinary protein complex (produced in the liver) of the pure-strain 129 (slot a) and C57 (b) controls and a control mixture (c), and the protein from the experimental mouse (d) in Fig. 4.*

Mosaic mouse #2, a female from an injected C57BL/6 blastocyst, was killed at 13 days of age. Although her coat was phenotypically like C57BL/6, she had internal mosaicism, as with some single-color allophenic mice (Mintz, 1970a, 1971a, 1974). With the glucosephosphate isomerase isozymic strain marker in starch gel analyses (Fig. 7), her blood cells were all C57BL/6, but she must have had many cells derived from the carcinoma cell implant in her thymus and kidneys. The morphologically female reproductive tract contained a minority

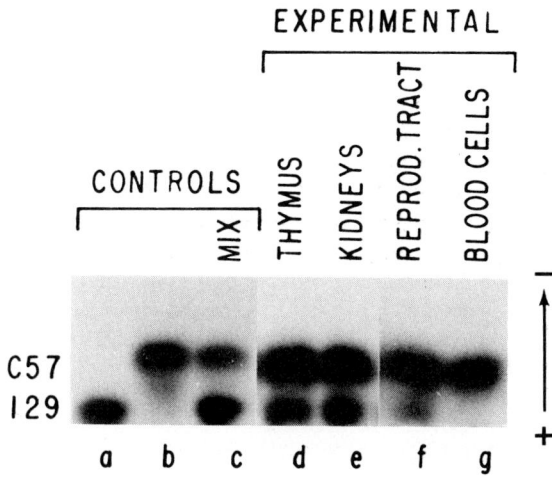

Figure 7. Glucosephosphate isomerase strain variants in starch gel electrophoresis of 129 (slot a), C57 (b) and mixed (c) controls, and in homogenates of specific tissues in female mosaic mouse #2 from a core-cell-injected C57BL/6 blastocyst.

of tumor-strain cells, which are known (Dunn and Stevens, 1970) to be of male (X/Y) sex chromosome type. This is consistent with earlier observations in allophenic mice, in which X/X↔X/Y cellular sex chromosome mosaicism can occur in animals with either sex phenotype, but functional gametes form only from germ cells compatible with the morphological sex (Mintz, 1969).

Mosaic mouse #3 was a male from an injected C57BL/6-*b/b* blastocyst and was killed at 3 days of age. With the glucosephosphate isomerase marker, his liver and kidneys comprised approximately one-third and his spleen approximately one-fifth of the 129-strain type. Other tissues examined (blood cells, brain, heart, lungs, thymus, stomach, intestines, pancreas, reproductive tract, and muscles) were all of the C57 type. Mosaicism restricted to a few tissues has previously been observed in some allophenic mice (Mintz and Palm, 1969).

In summary, the analyses of these three genetically

mosaic mice produced from blastocysts injected with embryonal carcinoma cells, obtained from cores of small-size embryoid bodies grown *in vivo*, have thus far shown substantial core-cell-derived cellular contributions in the following tissues: hair follicles (dermal papillae), melanocytes, red blood cells, circulating white blood cells (including all classes), antibody-producing plasma cells, spleen, hepatocytes, thymus, kidneys, and reproductive tract (not yet localized). The normal functioning of these tissues (apart from the absence of any visible tumors in them) is attested to by their formation of tissue-specific products whose provenance (partly or wholly) from tumor-derived cells is shown by their strain-specific allelic forms. The 129-type products that have been thus detected are: black melanin, pigment modifications of the phaeomelanin type, pigment modifications of the *steel* type (see the next section), adult hemoglobin of the *diffuse* type, 7S immunoglobulin allotypes of the G1 and of the G2a classes, the major urinary protein complex produced in hepatocytes, and the appropriate electrophoretic strain-variant of glucosephosphate isomerase in many tissues.

This list notably includes some tissues, such as liver, thymus, and kidney, never observed in mouse teratomas (Stevens and Hummel, 1957; Stevens, 1975, personal communication). In addition, although hairs are seen in 129-strain teratomas, no *agouti*-type hairs have been reported. Nor have antibody-producing cells been identified. Thus, teratocarcinoma stem cells have the inherent capacity to give rise to differentiated tissues that they are not able to form in the tumors. A possible reason for the absence of certain kinds of tissues in the tumors is that their development may depend on inductive (Grobstein, 1954) or other interactions between components that fail to become properly associated in the chaotically arranged growths.

According to conventional embryological classifications, the tumor-derived cells in our mosaic animals satisfy the criterion for multipotentiality, since they include contributions from all three germ-layers. This criterion continues to be generally invoked in evaluating the tissue lineages in teratomas or in teratocarcinoma cell cultures, as an index of multipotentiality (or lack of it) in the particular source of embryonal carcinoma cells from which the tumors or cultures were obtained.

It seems appropriate at this Symposium to open for consideration the question whether the classical germ-layer dogma is a useful basis for analyzing differentiation, especially in teratomas. As Oppenheimer (1940) has pointed out in a scholarly historical survey and critique entitled "The

Non-Specificity of the Germ-Layers", the germ-layer theory is based largely on topographical appearances; the so-called germ layers themselves have been shown to have much wider capacities for differentiation, depending on the circumstances, than those with which they are usually credited. Germ layers do normally arise in certain locations in the embryo, probably for important reasons that need to be understood; and they do normally undergo an orderly progression of movements and interactions, ending in specific accomplishments. But they have been shown to be experimentally interconvertible, each being capable of giving rise, in changed locations or conditions, to numerous structures usually formed from the others (Oppenheimer, 1940). It is therefore questionable whether any meaningful conclusions about germ-layer-related cell lineages in differentiation can be drawn in the context of the highly anomalous organization of teratomas. Even in the normal organization of the embryo itself, if differences were found among early germ-layers (e.g., in their molecular profiles), these would not necessarily be relevant to subsequent diversification. In later stages, any shared biochemical or other properties of differentiating or mature tissues would also not necessarily reflect their origin from a given germ-layer; if an 'ectoderm', 'endoderm,' or 'mesoderm' molecular phenotype existed in early stages, it would be unlikely to persist.

A less ambiguous approach than reliance on germ-layer theory would be to rely upon available experimental evidence, or to obtain new evidence, of specifically shared, or independent, developmental origins dating from inceptions of determination of specific tissues in teratomas or in embryos. For example, there is considerable experimental evidence (reviewed by Mintz, 1974) that erythrocytes and leukocytes, including lymphocytes, have a shared ancestry from more generalized hematopoietic stem cells; therefore, the 129-strain cells of these types in mosaic mouse #1 could all have diverged from a shared earlier cell or cell pool. Melanoblasts and hair follicle cells, on the other hand, are known to arise independently of each other (Mintz, 1974) and are unrelated to hematopoietic cell derivations. Hepatocyte origin is also unrelated to the others just named. In the other mosaic animals, it is not known which were the 129-strain cells in spleen and thymus; they may have been in the non-hematopoietically-derived supporting tissues. The 129-strain cells in kidneys and reproductive tract are also likely to have arisen separately from the contributions discussed above. Most tissues of mouse #1 (and 79 other animals in the series) are not yet analyzed. But these data already indicate that it is highly unlikely that each of the 5 core cells injected per blastocyst was already committed along a separate path of differentiation. Single core-cell injections would confirm this conclusion; these are being carried out (Illmensee and Mintz, experiments in progress).

The results appear strongly to support the probability that embryonal carcinoma core cells from OTT 6050 embryoid bodies always grown *in vivo* are developmentally totipotent. This probability refers to development of tissues in the embryo; we have not yet ascertained whether core cells can differentiate into extraembryonic membranes. In addition to their differentiation into the many tissues thus far identified in mosaic mice, we have also observed that core cells from the same source form other tissues (e.g., cartilage, muscle, nerve, epithelia) in solid tumors produced when cores are inoculated subcutaneously in syngeneic mice. Core cells also undergo myogenesis *in vitro* (Fig. 9). The capacity of these cells to differentiate into substantial portions of healthy normal mice is consistent with their euploid chromosome number.

Inasmuch as the conclusion that certain cells possess full capacity for normal development is based on an operational test, it cannot be generalized to other sources of embryonal carcinoma cells, each of which must be tested separately. In the only previous report of such tests, Brinster (1974) injected cells from completely dissociated embryoid bodies (cores plus rinds) into blastocysts of a random-bred albino strain, in which albinism was the only genetic marker distinguishing the embryo from the tumor strain. He obtained a single mouse (out of 60 survivors) with a few thin stripes of pigmented hairs on an albino background. However, the limited detectable contribution of 129-strain cells, and the low frequency of mosaicism, did not provide evidence for totipotency; it left open the possibility that an injected cell had simply already been determined as a melanoblast. The fact that the pigmented hairs were also *agouti*, as in 129-strain mice, did not prove that the hair follicle cells were also contributed by that strain. The blastocyst may have carried the *agouti* gene, hidden by albinism, and any pigment cells entering *agouti* follicles would produce *agouti* hairs (Silvers and Russell, 1955; Mintz, 1970a). The occurrence of melanoblast mosaicism was nevertheless encouraging for further *in vivo* investigation of developmental capacities of cells from embryoid bodies.

GENETIC ANALYSES OF CORE CELLS IN EMBRYOS

As stated in the introductory remarks, a long-range aim of the program in this laboratory is to cycle mutagenized core cells into differentiating mouse embryos, via blastocyst injections, in order to analyze genetic control mechanisms in differentiation. Three modes of analysis will be involved: observations of 'differentiation' phenotypes in the organism; biochemical and other analyses of mutant clones, before injection and after differentiation; and, if possible, recombinational analyses and genetic mapping through the germ line in progeny tests. There are at least preliminary indications that this sort of approach is a promising one.

The OTT 6050 teratoma was experimentally induced in 1967 by Stevens (1970), by transferring a 6-day embryo beneath the testis capsule, where it became disorganized and formed a solid transplantable teratoma. The tumor was then converted to the embryoid body ascites form by injecting minced fragments into the body cavity. The tumor strain of origin is highly inbred and is presumably homozygous at all loci, except for forced heterozygosity at the *steel* (Sl^J) locus: some animals are heterozygous $(Sl^J/+)$ and others wild-type $(+/+)$. (The *steel* homozygote dies prenatally.) The publication describing the origin of this tumor (Stevens, 1970) made no mention that either of the parents carried the *steel* gene in the particular experimental mating that was set up for production of the 6-day embryo. It was therefore completely unexpected when mosaic mouse #1 (Fig. 4) was seen to have the *steel* $(Sl^J/+)$ phenotype. This suggested, retrospectively, that at least one of the original tumor-embryo parents was $Sl^J/+$. (The *steel* coat is chiefly characterized by color dilution [Sarvella and Russell, 1956].) We therefore contacted Dr. L. C. Stevens who was, fortunately, able to trace in his 1967 records the details of the parentage of that 6-day embryo: its mother was indeed $Sl^J/+$, its father $+/+$ (Stevens, 1975, personal communication). Thus, the cycling of embryonal carcinoma core cells into differentiating mouse embryos has already enabled an unsuspected cryptic gene to be detected through its production in the soma of a differentiated organismic phenotype. This expression of a mutant gene provides a model for the experiments planned with mutant core cell injections into blastocysts.

Formation of other tissues (e.g., liver) in the mosaic animals, and of specialized proteins (e.g., immunoglobulins) whose molecular composition can be analyzed also demonstrates that it should be possible in this experimental system to detect the orderly expressions and products of many genes capable of influencing differentiation of somatic tissues in the embryo, but incapable of expressing themselves in the tumors or in cultures. Such somatic expressions would be highly useful experimental tools even if the mutant genes could not be analyzed in the germ line.

What, then, of the germ line in core-cell-derived mosaic mice? Experiments with allophenic mice have shown that functional gametes develop only from the germ cells whose sex chromosome constitution conforms to the morphological sex of the animal (Mintz, 1969). The known X/Y constitution of the OTT 6050 tumor (Dunn and Stevens, 1970) therefore requires that the mosaic animal be a male, in order that any tumor-derived germ cells produce gametes. Male morphology could result either from the fortuitous use of an X/Y blastocyst, or from predominance of tumor-strain cells in the reproductive tract of a sex chromosome mosaic. We have verified that mosaic mouse #1, who is a male, is fertile. He has sired normal-appearing embryos. Whether his gametes are of 129 type, C57 type, or both, is currently a 'cliff-hanger' but will soon be known from genetic markers in the progeny of his many matings with C57BL/6-*b/b* females.

IMPLICATIONS FOR CARCINOGENESIS

Eight years and almost 200 transplant generations of a highly malignant tumor have separated the two pairs of parents that gave rise to our 'four-parent' mosaic animals. The first pair of parents produced the embryo experimentally converted to a tumor in 1967 (Stevens, 1970); the second pair contributed a blastocyst that was injected with the carcinoma cells.

The tumor has continued to be highly malignant; yet the mosaic mice, with many tumor-derived cells, appear normal and healthy. The ability of these embryonal carcinoma cells to form normally functioning adult tissues is a clear indication that their initial transformation to malignancy did not involve structural genetic change. Inception and maintenance of malignancy in the stem cells seems to have been caused by a disorganized cellular environment rather than by mutation. Introduction of the cancer cells into the normal organizational framework of a blastocyst has proved to be a sufficient basis for termination of uncontrolled proliferation of the malignant stem cells and their complete reversal to normalcy.

Pierce (1967) has proposed, in a broad sense, that carcinogenesis may be a result of altered developmental influences rather than of alteration of the genome and he has discussed this with particular reference to teratomas. Stevens

(1967) and Dunn and Stevens (1970) have pointed out that embryo-derived experimental teratomas, such as OTT 6050, might, theoretically, arise either after cells comparable in developmental potential to primordial germ cells had appeared in the graft and became neoplastic, or simply as a result of disruption of normal cellular interrelationships, so that inductive or other interactions were prevented, thus leading to maintenance of the undifferentiated state and neoplastic conversion of some embryo cells. From various lines of evidence, Stevens believed that the second explanation was correct. In the case of spontaneous teratocarcinogenesis, the process seems to start with spontaneous parthenogenesis of germ cells (in testicular teratomas, in the diploid state), giving rise to a population of undifferentiated embryonic cells which, as in the grafted embryos, become neoplastic due to abnormal cellular relationships. Thus, in both spontaneous and experimental teratomas, the actual neoplastic change was thought to occur in relatively undifferentiated embryonic cells rather than in the germ cells from which the disordered embryos developed.

The differentiation of tumor-derived mosaic mice offers strong evidence that these hypotheses are correct. It now seems entirely admissible that at least some other kinds of tumors, from later-stage stem cells of more specialized tissues, may have a comparable, non-mutational, origin, beginning with tissue disorganization and leading to changes in gene expression, hence alteration of stem cell differentiation. In previous studies of tumorigenesis in allophenic mice (Mintz, 1970b), the view was also expressed that changes in gene expression were primarily involved in carcinogenesis and that they led to maintenance of proliferative states more normal for cells of earlier stages.

DIFFERENTIATION *IN VITRO*: MYOGENESIS

Short-term cultures provide opportunities to observe, mutagenize, select, manipulate, or radioactively label core cells. Our earliest experiments with cultures were conducted with whole embryoid bodies, which were available in unlimited quantities (Gearhart and Mintz, 1974). Embryoid bodies of different size classes, but especially of the smallest size (less than 100 µm) were used, as these are enriched for the core-cell component and are unlikely to contain differentiating cells. When the embryoid bodies were grown in suspension, as they are *in vivo* in the ascites fluid, they failed to differentiate. When they were allowed to attach to the plastic culture dish, analogous to attachment to tissue surfaces on which solid tumors differentiate *in vivo*, differentiation occurred. Attachment of the cell surface to a substratum therefore plays some critical role in initiating developmental commitments. These cultures at first appeared highly heterogeneous, as others have also noted independently in embryoid body explants (Levine et al., 1974; Lehman et al., 1974). We are somewhat puzzled by the fact that the other laboratories have not observed, as we did, that the cultures soon showed very conspicuous myogenesis. This begins with formation of characteristic multinucleated myotubes and culminates in formation of a huge mat of muscle fibers. Early stages of these events, without the yolk sac cells, have also been confirmed in the present study of cores isolated from embryoid bodies (Figs. 8, 9).

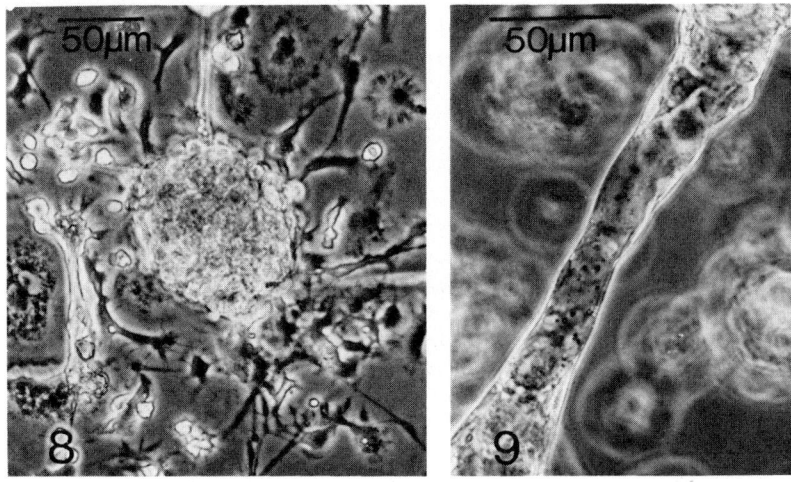

Figure 8. A core from an embryoid body attaching and spreading in culture. Phase contrast optics.
Figure 9. A multinucleated myotube formed in culture from a core explanted as in Fig. 8. Phase contrast optics.

Myogenesis is therefore the process most readily examined biochemically in these cultures, in comparison with known parameters of normal myogenesis. On the basis of the present report, showing totipotency of core cells in embryoid bodies, it can be said that both myogenic *determination* from uncommitted cells and myogenic *differentiation* must be occurring in the cultures. However, the muscle-phenotype parameters thus far examined in them probably relate chiefly to relatively late events, i.e., to differentiation rather than determination.

In the embryoid body cultures, myogenesis is not entirely normal: the fibers have a paucity of striations and are not contractile. Nevertheless, acetylcholinesterase activity increases, especially during myotube formation, as in normal myogenesis (Gearhart and Mintz, 1974, 1975). Myokinase activity does not increase appreciably. Creatine kinase activity rises during myotube formation and then drops abnormally. The fetal isozymic form of creatine kinase, also known to occur in brain, is expressed in the cultures, although well differentiated solid tumors taken from mice contain the adult muscle isozyme type, provided that skeletal muscle is demonstrably present in histological sections examined from nearby areas of the tumor. The results are consistent with the possibility that later stages of myogenesis may require coordinately regulated changes in gene expressions controlling these various functions. In another report, by Levine et al. (1974), in which explanted embryoid bodies were not observed to undergo myogenic differentiation, only the fetal or brain isozymic form of creatine kinase was found in the cultures or in solid tumors. This, and the rise in acetylcholinesterase activity, led them to believe that neurogenesis was being monitored in neuronal-like cells. However, absence of the adult muscle type of creatine kinase in their solid tumor samples could have been due to chance sampling of muscle-free tumors, which were not examined histologically. Moreover, the identity of morphologically 'neuronal-like' cells with processes, seen in these cultures, is less certain than that of the myotubes and muscle fibers. It is nevertheless possible that dissimilar avenues of differentiation, due either to different tumor sublines or culture conditions, prevailed in the two experiments.

RELATIONSHIP BETWEEN EMBRYONAL CARCINOMA CELLS AND EMBRYO CELLS

In the last aspect of our present studies to be considered here, we take up the question whether the apparently totipotent core cells of OTT 6050 embryoid bodies correspond to pre-blastocyst-stage cells, which are known, from other experimental evidence (reviewed by Mintz, 1974), to be totipotent, or to some later-stage cells for which there is not yet direct evidence of totipotency. There has perhaps been a widespread assumption that readily available teratocarcinoma cells would, in effect, conveniently provide us with a 'barrel' of cells equivalent to the less readily available blastomeres. But even preliminary explorations suggest that they are not similar.

One point, already mentioned, concerns undefined differences in their surface properties: although core cells stick very firmly to each other (Fig. 2), they do not adhere well to cleavage- or morula-stage cells.

Another line of evidence comes from comparisons of soluble proteins by acrylamide gel disc electrophoresis methods (Ornstein, 1964; Davis, 1964). Details of tissue handling, electrophoresis, staining in Coomassie Brilliant Blue, and microphotometric scanning will be presented elsewhere (Gearhart and Mintz, manuscript in preparation). In the gel scans in Fig. 10, protein profiles are being compared in 129-strain morulae (with about 10-16 cells each), cores, and small-size embryoid bodies (less than 100 μm); fetal mouse serum is also shown. The high peaks of the morulae represent

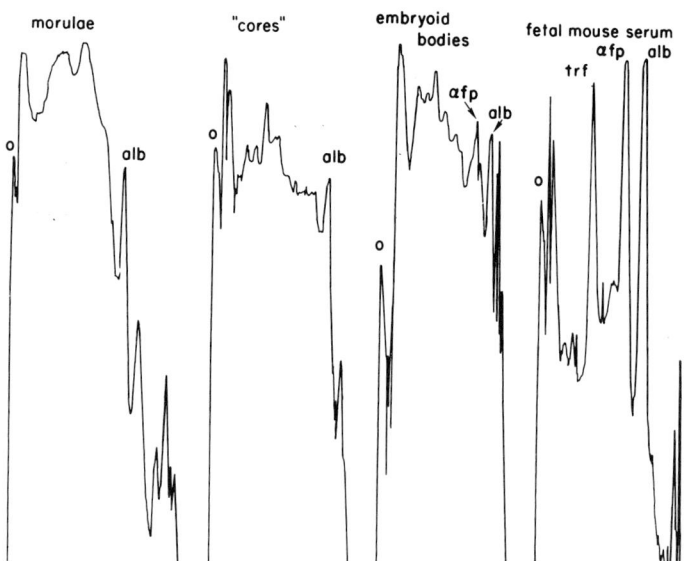

Figure 10. Scans of soluble proteins separated by acrylamide disc gel electrophoresis of mouse morulae, cores, and embryoid bodies. Fetal mouse serum is shown for comparison. α-fp, α-fetoprotein; alb, albumin; trf, transferrin; o is the origin.

fairly large amounts of protein and may be obscuring multiple components which may later prove to be resolvable. The method is of course less sensitive than some others for minor components. In addition to the tissues shown in Fig. 10, a 129-strain transplantable yolk sac carcinoma, received from Dr. G. B. Pierce, was analyzed. This carcinoma, which is not multipotential, was derived from the OTT 6050 ascites teratoma. It has a parietal yolk sac epithelium surrounding a hyaline secretion, as described for another such tumor (Pierce and Dixon, 1959).

Albumin is present in morulae, cores, whole embryoid bodies, and yolk sac carcinoma. In the morulae, this might conceivably be due to uptake from the maternal environment rather than synthesis in the embryo cells, inasmuch as Glass (1963) has detected uptake into preimplantation mouse embryos *in vivo* under experimental conditions.

α-Fetoprotein is an interesting glycoprotein with variant forms due in part to variations in sialic acid residues. It has been especially intensively studied in human material because it appears as a normal product of fetal yolk sac and liver, disappears in adults, and tends to reappear in hepatomas and teratomas and in the serum of patients with these tumors (Abelev, 1974). It has been under rigorous scrutiny as a possible early clinical indicator of hepatomas. α-Fetoprotein is present in yolk sac carcinoma and embryoid bodies, but not in detectable amounts in either cores or morulae (Fig. 10). Identification of α-fetoprotein from embryoid bodies and yolk sac carcinoma was further confirmed by co-electrophoresis with fetal mouse serum, the use of rabbit anti-mouse-α-fetoprotein serum (prepared by Dr. Bruce Smith of this Institute), and stainability with the periodic acid-Schiff reagent. These results support the work of Abelev (1974) and of Engelhardt et al. (1973). Based on *in situ* immunofluorescence in conjunction with other tests, they concluded that α-fetoprotein is specifically produced by the endodermal and yolk sac cells of teratomas, rather than by their embryonal carcinoma cells. α-Fetoprotein has been reported to originate in the embryonal carcinoma cells (Kahan and Levine, 1971) in an *in vitro* clonal line of teratocarcinoma cells from the transplantable SV 6965 tumor which arose as a spontaneous testicular teratoma in an F_1(A/He x 129/Sv) hybrid. The discrepancy could be accounted for if some differentiation of yolk sac cells from multipotential cells were taking place in the cultures, as clonal origin of a multipotential cell line does not preclude undetected differentiation of some cells in the cultures of that cell line.

Except for absence or undetectably low levels of α-fetoprotein in cores, the electrophoretic protein profiles in general appear very similar in cores, embryoid bodies, and parietal yolk sac carcinoma. However, the morulae seem to have a number of differences from core cells even though, as we now know, both kinds of cells are developmentally totipotent.

There are other indications that embryonal carcinoma cells appear not to be equatable to morula cells. Embryonal carcinoma cells are very rich in alkaline phosphatase (Damjanov et al., 1971; Bernstine et al., 1973), unlike preimplantation embryos but similar to ecto-mesodermal cells of the postimplantation egg cylinder stage (Damjanov et al., 1971). Ultrastructural comparisons also point to this similarity (see Damjanov and Solter, 1975).

It remains uncertain which phenotypic parameters in embryonal carcinoma cells or embryo cells are relevant to developmental totipotency. For example, embryonal carcinoma cells from all sources resemble each other; yet those from some sources are not multipotential and give rise to malignant tumors with no differentiated tissues.

In the case of the OTT 6050 tumor studied here, it would not be surprising if the embryonal carcinoma cells corresponded to some postimplantation-stage cells, as the tumor itself arose from a 6-day-old embryo. There is as yet no direct evidence on the duration of a totipotent cell population in the mouse embryo. Stevens (1970) has found that embryos older than 6 1/2 days could no longer form teratocarcinomas when transplanted to an ectopic site. Mintz (1970a, 1974) has inferred from indirect clonal analyses in $vivo$ that the various cell lineages under study had all originated as specific determinations in the day 5-7 period of embryonic life. Both conclusions, from quite dissimilar investigations, are completely consistent: they tend to suggest that some totipotent cells may in fact persist through day 6 in mouse embryos, but not appreciably beyond. Embryonal carcinoma cells may therefore correspond to still-totipotent stem cells in postimplantation embryos of approximately 6 days of embryonic age. In teratomas experimentally induced by grafting younger embryos to ectopic sites, embryo development may continue, even if imperfectly, until stem cells of the age appropriate for neoplastic conversion are present.

ACKNOWLEDGEMENT

These investigations were supported by United States Public Health Service grants Nos. HD-01646, CA-06927, and RR-05539, and by an appropriation from the Commonwealth of Pennsylvania.

REFERENCES

Abelev, G. I. (1974). α-Fetoprotein as a marker of embryo-specific differentiations in normal and tumor tissues. Transplant. Rev. 20, 3.

Bernstine, E. G., Hooper, M. L., Grandchamp, S., Ephrussi, B. (1973). Alkaline phosphatase activity in mouse teratoma. Proc. Nat. Acad. Sci. U. S. 70, 3899.

Bosma, M. J., Bosma, G. C. (1974). Congenic mouse strains: The expression of a hidden immunoglobulin allotype in a congenic partner strain of BALB/c mice. J. Exp. Med. 139, 512.

Brinster, R. L. (1974). The effect of cells transferred into the mouse blastocyst on subsequent development. J. Exp. Med. 140, 1049.

Condamine, H., Custer, R. P., Mintz, B. (1971). Pure-strain and genetically mosaic liver tumors histochemically identified with the β-glucuronidase marker in allophenic mice. Proc. Nat. Acad. Sci. U. S. 68, 2032.

Damjanov, I., Solter, D., Skreb, N. (1971). Enzyme histochemistry of experimental embryo-derived teratocarcinomas. Zeitschr. Krebsforsch. 76, 249.

Damjanov, I., Solter, D. (1975). Ultrastructure of murine teratocarcinomas. This volume, p. 209.

Davis, B. J. (1964). Disc electrophoresis - II. Method and application to human serum proteins. Ann. N. Y. Acad. Sci. 121, 404.

Dunn, G. R., Stevens, L. C. (1970). Determination of sex of teratomas derived from early mouse embryos. J. Nat. Cancer Inst. 44, 99.

Engelhardt, N. V., Poltoranina, V. S., Yazova, A. K. (1973). Localization of alpha-fetoprotein in transplantable murine teratocarcinomas. Int. J. Cancer 11, 448.

Evans, M. J. (1972). The isolation and properties of a clonal tissue culture strain of pluripotent mouse teratoma cells. J. Embryol. Exp. Morph. 28, 163.

Finlayson, J. S., Asofsky, R., Potter, M., Runner, C. C. (1965). Major urinary protein complex of normal mice: Origin. Science 149, 981.

Gardner, R. L. (1968). Mouse chimaeras obtained by the injection of cells into the blastocyst. Nature (London) 220, 596.

Gearhart, J. D., Mintz, B. (1972). Clonal origins of somites and their muscle derivatives: Evidence from allophenic mice. Develop. Biol. 29, 27.

Gearhart, J. D., Mintz, B. (1974). Contact-mediated myogenesis and increased acetylcholinesterase activity in primary cultures of mouse teratocarcinoma cells. Proc. Nat. Acad. Sci. U. S. 71, 1734.

Gearhart, J. D., Mintz, B. (1975). Creatine kinase, myokinase, and acetylcholinesterase activities in muscle-forming primary cultures of mouse teratocarcinoma cells. Cell 6, 61.

Glass, L. E. (1963). Transfer of native and foreign serum antigens to oviducal mouse eggs. Amer. Zool. 3, 135.

Grobstein, G. (1954). Tissue interaction in the morphogenesis of mouse embryonic rudiments *in vitro*. In "Aspects of Synthesis and Order in Growth". (Ed. D. Rudnick). Princeton University Press, p. 233.

Jakob, H., Boon, T., Gaillard, J., Nicolas, J.-F., Jacob, F. (1973). Teratocarcinome de la souris: Isolement, culture et proprietes de cellules a potentialites multiples. Ann. Microbiol. (Inst. Pasteur) 124 B, 269.

Kahan, B. W., Ephrussi, B. (1970). Developmental potentialities of clonal *in vitro* cultures of mouse testicular teratoma. J. Nat. Cancer Inst. 44, 1015.

Kahan, B., Levine, L. (1971). The occurrence of a serum fetal α_1 protein in developing mice and murine hepatomas and teratomas. Cancer Res. 31, 930.

Kleinsmith, L. J., Pierce, G. B., Jr. (1964). Multipotentiality of single embryonal carcinoma cells. Cancer Res. 24, 1544.

Lehman, J. M., Speers, W. C., Swartzendruber, D. E., Pierce, G. B. (1974). Neoplastic differentiation: Characteristics of cell lines derived from a murine teratocarcinoma. J. Cell. Physiol. 84, 13.

Levine, A. J., Torosian, M., Sarokhan, A. J., Teresky, A. K. (1974). Biochemical criteria for the *in vitro* differentiation of embryoid bodies produced by a transplantable teratoma of mice. The production of acetylcholine esterase and creatine phosphokinase by teratoma cells. J. Cell. Physiol. 84, 311.

Lin, T. P. (1966). Microinjection of mouse eggs. Science 151, 333.

Mintz, B. (1967). Gene control of mammalian pigmentary differentiation. I. Clonal origin of melanocytes. Proc. Nat. Acad. Sci. U. S. 58, 344.

Mintz, B. (1969). Developmental mechanisms found in allophenic mice with sex chromosomal and pigmentary mosaicism. In "Birth Defects: Original Article Ser. 5" (Eds. D. Bergsma, V. McKusick). National Foundation, New York, p. 11.

Mintz, B. (1970a). Gene expression in allophenic mice. In "Symp. Int. Soc. Cell Biol." (Ed. H. Padykula). Academic Press, New York, Vol. 9, p. 15.

Mintz, B. (1970b). Neoplasia and gene activity in allophenic mice. In "Genetic Concepts and Neoplasia". Williams and Wilkins, Baltimore, p. 477.
Mintz, B. (1971a). Clonal basis of mammalian differentiation. In "Symp. Soc. Exp. Biol". (Eds. D. D. Davies, M. Balls) Cambridge University Press, p. 345.
Mintz, B. (1971b). Allophenic mice of multi-embryo origin. In "Methods in Mammalian Embryology" (Ed. J. Daniel, Jr.). Freeman, San Francisco, p. 186.
Mintz, B. (1974). Gene control of mammalian differentiation. Ann. Rev. Genet. 8, 411.
Mintz, B., Illmensee, K. (1975). Normal genetically mosaic mice produced from malignant teratocarcinoma cells. Proc. Nat. Acad. Sci. U. S., in press.
Mintz, B., Palm, J. (1969). Gene control of hematopoiesis. I. Erythrocyte mosaicism and permanent immunological tolerance in allophenic mice. J. Exp. Med. 129, 1013.
Nicolas, J.-F., Dubois, P., Jakob, H., Gaillard, J., Jacob, F. (1975). Teratocarcinome de la souris: Differenciation en culture d'une lignee de cellules primitives a potentialites multiples. Ann. Microbiol. 126 A, 3.
Oppenheimer, J. M. (1940). The non-specificity of the germ layers. Quart. Rev. Biol. 15, 1.
Ornstein, L. (1964). Disc electrophoresis - I. Background and theory. Ann. N. Y. Acad. Sci. 121, 321.
Pierce, G. B., Jr. (1967). Teratocarcinoma: Model for a developmental concept of cancer. Curr. Top. Develop. Biol. 2, 223.
Pierce, G. B., Dixon, F. J., Jr. (1959). Testicular teratomas. II. Teratocarcinoma as an ascitic tumor. Cancer 12, 584.
Ranney, H. M., Smith, G. M., Gluecksohn-Waelsch, S. (1960). Haemoglobin differences in inbred strains of mice. Nature (London) 188, 212.
Sarvella, P. A., Russell, L. B. (1956). Steel, a new dominant gene in the house mouse. J. Hered. 47, 123.
Silvers, W. K., Russell, E. S. (1955). An experimental approach to action of genes at the agouti locus in the mouse. J. Exp. Zool. 130, 199.
Stevens, L. C. (1967). The biology of teratomas. Advan. Morph. 6, 1.
Stevens, L. C. (1970). The development of transplantable teratocarcinomas from intratesticular grafts of pre- and postimplantation mouse embryos. Develop. Biol. 21, 364.
Stevens, L. C., Hummel, K. P. (1957). A description of spontaneous congenital testicular teratomas in strain

129 mice. J. Nat. Cancer Inst. 18, 719.

Notes added in proof. Further analyses of cases from blastocysts injected with embryonal carcinoma cells have revealed carcinoma-derived cells in the normal placental tissues of 11-14-day fetuses. In addition, the progeny tests of mosaic mouse #1 have now shown that his sperms are entirely or largely derived from the carcinoma cells; some sperms have also transmitted the *steel* gene.

EXPERIMENTAL TERATOMAS IN RATS

NIKOLA SKREB AND ANTON SVAJGER

Department of Biology and Department of Histology
and Embryology, Faculty of Medicine, Zagreb,
Yugoslavia

INTRODUCTION

To review the data on teratomas in rats is a thankless task. This is mainly due to two reasons: first, spontaneous teratomas have never been reported in rats, and second, experimental teratomas in rats have been described so far as a result of experiments dealing with problems of cell differentiation in ontogenesis rather than those concerned with the induction of tumours. As the rationale is different in these two approaches, it is rather difficult to interpret the available data in terms of the problem of experimental oncogenesis. Even the term teratoma has been very rarely used by authors working with experimental teratomas in rats. Such authors comported themselves like Moliere's "Bourgeois gentilhomme" who had spoken in prose all his life without having been aware of it.

Nevertheless, experiments with the transplantation of rat embryonic shields to different extrauterine sites, and also those with explantation *in vitro*, can probably shed some light on problems of the induction of tumours. To attain this, one needs to focus attention on the following pertinent questions:

1) Does the developmental capacity of early embryonic cells depend on the developmental stage at the moment of isolation, and on the type of the extrauterine environment?

2) Is it possible to find an extrauterine environment which would stimulate growth of embryonic cells without permitting their further differentiation (i.e. which would induce teratocarcinogenesis)?

3) Does the interaction of embryonic cells with various extrinsic factors follow the rule of an 'all or none' reaction or rather that of 'more or less' differentiation?

The first substantial contribution to the solution of the above listed problems was probably the paper of Nicholas and Rudnick (1933). They transplanted 6- to 11-day rat embryos (with and without extra-embryonic membranes) onto chick chorioallantoic membrane. Only few transplants became incorporated into the membrane and gave rise to differentiated tissues, such as cartilage, nervous tissue and intestine. Subsequently, Nicholas (1942) investigated the differentiation of pre- and post-implantation rat embryos transferred under the kidney capsule. The transferred cleaving eggs sometimes developed into teratomas consisting of tissue derivatives of all three germ layers. The post-implantation stages (egg-cylinders), transferred before and after mesoderm formation, gave more positive results, and mesodermal derivatives were predominant constituents of the resulting teratomas. For some unknown reason, but most probably because of immunological intolerance, all grafts were ultimately resorbed. The results of more recent investigations (Fawcett, 1950; Kirby, 1962) have not confirmed Nicholas' results concerning 2- to 4-cell eggs. Kirby (1962) transplanted 120, 2- to 4-cell rat eggs under the kidney capsule. About 75% of them were successfully incorporated, but contrary to the results obtained by Nicholas, they developed like the 2- and 4-cell mouse eggs, i.e. the embryonic shield proper did not differentiate, whereas the trophoblast flourished and actively invaded the host tissue.

TRANSPLANTATION ONTO CHICK CHORIOALLANTOIC MEMBRANES

Our attempts to repeat Nicholas and Rudnick's experiments and to get cytodifferentiation of the rat egg-cylinder transplanted on chick chorioallantoic membrane (CAM) have failed. Although, in particular experimental series, we got about 50% of the grafts incorporated successfully, the final differentiation of tissues was poor. The restricted 6-day period of keeping rat embryo heterografts in the chick CAM

ABBREVIATIONS AND TERMINOLOGY

CAM - chorioallantoic membrane.

The day of observation of the sperm plug is considered day zero of pregnancy.

was obviously too short to allow any marked cyto- and histo-differentiation. Even the re-transplantation of the grafted material after 5 days did not improve the final result. The non-specific reactions of the CAM to the presence of heterografts have proved to be more interesting than the results of differentiation of the grafted rat embryonic material. Therefore, transplantation on chick CAM as a method for testing the developmental capacities of early rat embryonic cells in an extrauterine environment was abandoned.

TRANSPLANTATION UNDER THE KIDNEY CAPSULE

Whole Embryonic Shields

We have also tested the developmental capacities of the rat egg-cylinder isolated at different developmental stages (7 1/2- and 9-day old rat embryo), and transplanted under the kidney capsule of isogeneic adult male animals (Skreb et al., 1971). Random-bred and inbred Fischer rats were used in the experiment. The majority of grafts were analysed histologically after 15 or 30 days, but some of them were left under the kidney capsule for 2, 4 or 6 months. In general, our results have confirmed those of Nicholas (1942) with post-implantation rat embryonic shields. The resulting teratomas consisted of tissue-derivatives of all three definitive germ layers. Differentiation was the same in embryos transplanted at 7 1/2 and 9 days, both in random-bred and inbred rats.

As a rule, the teratomas were composed of well-differentiated tissues. Some of them (neural tissue, skeletal muscle, bone, and partly cartilage) were present in chaotic mixtures (Fig. 1). Epithelial tissues, however, most frequently appeared in organotypic combinations with tissues of mesodermal origin. The ciliated pseudostratified columnar epithelium of the respiratory tube was often surrounded by non-ossifying hyaline cartilage. On the other hand, tubes or cysts lined with epithelium of the oesophagus, stomach or intestine were usually surrounded by two layers of smooth muscle, which sometimes even contained small intramural ganglia (Fig. 2). Regionally specific characteristics of the gut were well expressed (histologically distinct surface and glandular epithelium of the oesophagus, stomach, small and large intestine). If, at the head-fold stage, circumscript areas of the embryonic shield were transplanted, regionally specific derivatives of particular germ layers predominated in the resulting teratomas. The anterior part of the embryonic shield (corresponding to the extension of the neural groove) gave rise to head structures (brain, choroid plexus, vibrissae, teeth), as well as to derivatives of the foregut (tongue, oesophagus, thymus, thyroid). The posterior part of the shield (corresponding to the extension of the primitive streak), however, gave rise to endodermal derivatives of the mid- and hindgut (small and large intestine, urogenital sinus, prostatic complex) (Svajger and Levak-Svajger, 1974).

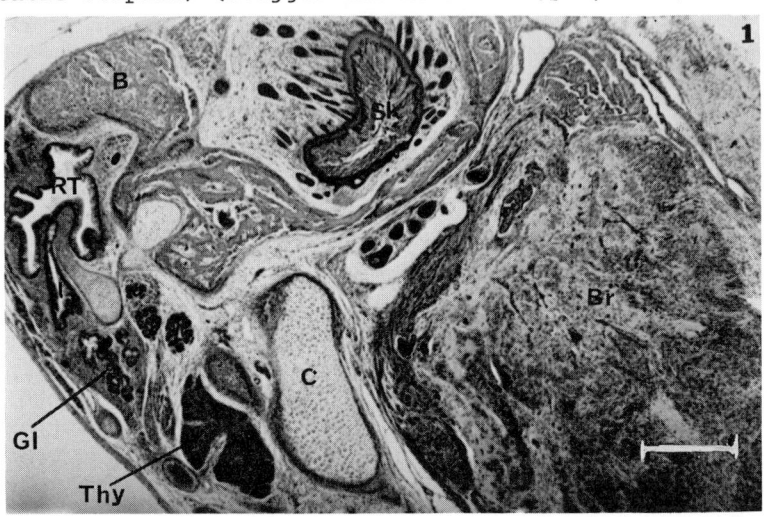

Figure 1. Histological structure of a teratoma derived from a renal homograft of the anterior part of a head-fold stage rat embryonic shield. B, bone; Br, brain; C, cartilage; Gl, glands; RT, respiratory tube; Sk, skin; Thy, thymus. Scale = 0.3 mm. (From: A. Svajger and B. Levak-Svajger (1974) J. Embryol. Exp. Morph. 32, 461, by permission of the Company of Biologists Ltd.)

It must be pointed out that hollow epithelial organs (respiratory tube, intestine), which are passable tubes *in situ*, develop in teratomas in the form of closed cysts. This feature (lack of drainage) results in the accumulation of secretion, increased inner pressure, and consequently, the flattening of the lining epithelium, accompanied by loss of the elaborate organ-specific epithelial pattern. These epithelial cells must not be considered as undifferentiated or dedifferentiated ones, for it has been shown that they reassume the capacity to differentiate into the organ-specific mucosa if the cysts are emptied by incision (Usadel et al., 1970).

Histological findings suggest that some teratomas undergo partial necrosis in the course of their development. The neural tissue seems to be especially susceptible. The best example was a graft consisting exclusively of a well differentiated intestine (typical surface and glandular epithelium, two layers of smooth muscle fibers). Intramural ganglia between the two muscular layers (Auerbach's myenteric plexus) were present, although neural tissue outside the intestine was necrotic.

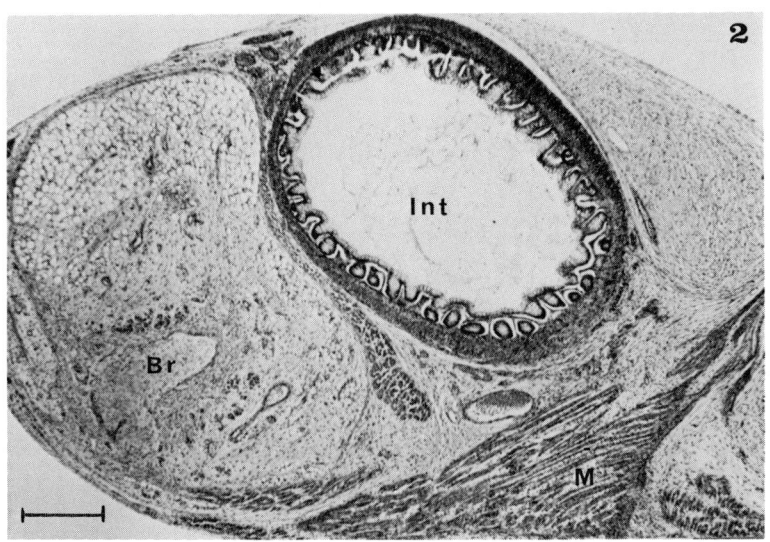

Figure 2. Organotypic combination of tissues (complete structure of the intestinal wall) in a teratoma derived from a renal homograft of the posterior part of a head-fold stage rat embryonic shield. Br, brain; Int, intestine; M, skeletal muscle. Scale = 0.3 mm. (From: A. Svajger and B. Levak-Svajger (1974) J. Embryol. Exp. Morph. 32, 461, by permission of the Company of Biologists Ltd.)

As far as the growth-rate of egg-cylinders transplanted under the kidney capsule is concerned, some statistically significant differences between different strain-characterized graft-host combinations have been observed (Table 1). In general, the random-bred rat embryos grew better than the Fischer-strain ones. The Fischer-strain embryos grew better in the random-bred host than in the isogeneic Fischer-strain host. On the contrary, the increase in weight of the random-bred embryo grafts was not significantly different in the random-bred and in the Fischer-strain host. These results can be interpreted as suggesting that factors other than immunological ones are responsible for the differences in growth of the embryonic grafts (Skreb et al., 1971).

TABLE 1

WEIGHTS OF EGG-CYLINDER GRAFTS
TRANSPLANTED UNDER THE KIDNEY CAPSULE

Series	Strain	Experimental Period (days)	No. Successful Grafts	Weight[b] (mg)	Weight of Contralateral Kidney[b] (mg)
1	F → F[a]	15	10	31 ± 6[c]	574 ± 20
2	R → R	15	10	135 ± 20[d]	885 ± 30
3	F → R	15	14	73 ± 7[c]	721 ± 20
4	R → F	15	10	95 ± 15[d]	716 ± 30
5	F → F	60	6	1587 ± 690	965 ± 20
6	F → F	120	6	3374 ± 950	1032 ± 45
7	F → F	180	10	9208 ± 3080	1271 ± 50

[a]F → F signifies a Fischer strain embryo transplanted to a Fischer strain rat, R → F signifies a random-bred embryo transplanted to a Fischer strain rat, etc.

[b]± standard error of the mean.

[c]Difference between series 1 and 3: $t = 3.59$, $P < 0.01$.

[d]Difference between series 2 and 4: $t = 1.59$, $P < 0.1$.

In a single case of a 9-day Fischer-strain rat egg-clyinder cultivated under the kidney capsule for 6 months (out of 22 cases) a very large tumour, weighing about 31 g (ca. 30 x the average weight of the adult rat kidney) was obtained. It was histologically characterized by a large necrotic center, very few well differentiated tissues, and large masses of apparently undifferentiated cells (Fig. 3). Some isolated areas, which were histologically similar to the murine yolk sac carcinoma, were also present. In spite of the fact that this tumour was histologically very similar to the teratocarcinoma obtained in grafts of mouse embryos, it is very difficult to interpret it as a true teratocarcinoma for its capability for re-transplantation (one of the principal properties of the teratocarcinoma, see Damjanov and Solter, 1974) was not tested, because the problem of teratocarcinogenesis was not the subject of this investigation.

Separate Germ Layers

In order to analyze in more detail the developmental capacities of isolated parts of early rat embryos, we have

worked out in our laboratory a technique of separation of
single germ layers from rat embryonic shields (Levak-Svajger
et al., 1969). Some of the results obtained by using this
technique are also relevant to the problem of experimental
induction of teratomas (Levak-Svajger and Svajger, 1971, 1974,
and unpublished results). In all developmental stages tested
thus far (the pre-primitive streak, the early primitive streak
and the head-fold stage) the isolated endoderm does not differentiate at all if grafted alone under the kidney capsule

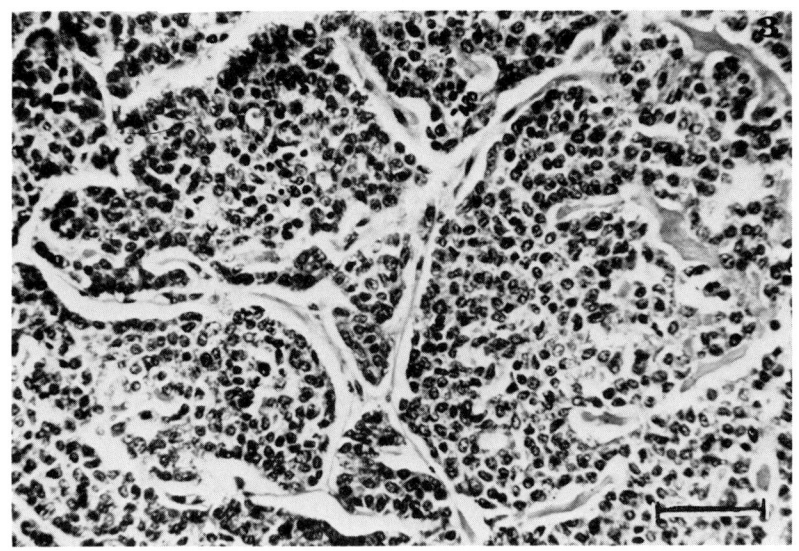

*Figure 3. Lobules of undifferentiated cells from a large
tumour derived from a head-fold stage rat embryonic shield
after 6 months under the kidney capsule. Scale = 0.05 mm.*

of isogeneic adult rats. However, if isolated at the head-
fold stage and grafted together with its adjacent mesoderm,
it differentiates into various, histologically well-defined
derivatives of the primitive gut. The embryonic mesoderm
isolated from the head-fold to the 12-somite-embryo, gives
rise almost exclusively to brown adipose tissue. Embryonic
ectoderm, isolated and grafted at the pre-primitive streak
and the early primitive streak stage, gives rise to teratomas
which consist of mature tissues, derivatives of all three
definitive germ layers. The regular presence of typical endodermal derivatives (i.e. derivatives of the primitive gut) in
grafts of early embryonic ectoderm is especially significant
with regard to the problem of origin of definitive germ

layers, and also to the conceptual consideration of the germ-layer theory in general. When isolated and transplanted toward the end of the period of gastrulation (the head-fold egg-cylinder), the embryonic ectoderm differentiates only into tissue derivatives of the definitive ectoderm and mesoderm.

One may conclude from our experiments with the transplantation of isolated germ layers that the histological composition of the resulting teratomas is strongly dependent on the developmental stage of the egg-cylinder from which the germ layers are isolated. One must especially point out that all these tumours consist exclusively of normal, more or less mature tissues (dependent on the time of development in the host). Undifferentiated cells, like those which are usually found in murine teratocarcinomas, have never been observed.

We have made a further attempt to study the behaviour of early rat egg-cylinder at different time intervals after grafting under the kidney capsule. Up to seven days after transplantation, some isolated clusters of undifferentiated embryonic cells could be found within the grafts. Two weeks after transplantation, undifferentiated cells were no longer observed in rat embryo grafts. In all mouse egg-cylinders grafted under the kidney capsule of isogeneic hosts, however, groups of undifferentiated cells like those usually found in murine teratocarcinomas have been observed when grafts were examined two weeks after transplantation (Damjanov and Solter, 1974). It seems that some as yet unknown environmental factors present under the kidney capsule of the rat favor the differentiation of all early rat embryonic cells. Another possible explanation concerns some host-related factors, possibly of an immunologic nature, which govern and regulate the regression of undifferentiated embryonic cells. On the basis of existing data it is difficult to decide between these, or possibly some other, interpretation of the described feature.

TRANSPLANTATION INTO THE ANTERIOR CHAMBER OF THE EYE

The anterior chamber of the eye is known as a suitable environment for testing the developmental capacities of mammalian embryonic shields (Grobstein, 1951). We transplanted to this site the pre-primitive streak (7 1/2 days), the early primitive streak (8 1/2 days) and the head-fold stage (9 1/2 days) rat embryonic shields (Levak-Svajger and Skreb, 1965). The resulting teratomas did not differ substantially from those obtained after transfer of the same stages under the kidney capsule. The only marked differences were: a) a lower percentage of successfully incorporated and differentiated grafts, and b) poor differentiation of mesodermal derivatives (cartilage, bone, muscle) in teratomas derived from embryonic shields before the onset of mesoderm formation. The following explanation of this difference seems to be the most plausible: by the act of transfer under the kidney capsule, the embryonic shield is immediately placed between two well vascularized tissues, the kidney capsule and the kidney parenchyma. Blood capillaries bud into the graft from both sides, the graft being thus vascularized immediately, without an initial period of penetration into the host tissue. The connection between the graft and the host tissues remains only vascular. This statement seems to be confirmed by the experience that the resulting teratomas can be, as a rule, easily and 'cleanly' detached from the surface of the renal cortex.

After transfer into the anterior chamber of the eye, however, the embryonic shield first floats freely in the aqueous humour, and has to come into contact with, and penetrate into, the iris in order to establish vascular connection with the host. Perhaps the aqueous humour cannot provide a young embryo with the nutrients necessary for the realization of all its developmental capacities. Later on, when incorporation and vascularization is established, the implants are no longer able in all cases to regain their lost developmental competence. This significant difference in the initial behaviour of grafts in different ectopic sites might well be considered as an explanation of differences in their capacities for differentiation.

The following experiment has helped us to better understand the significance of good vascularization for the differentiation of transplanted embryos (Skreb and Levak, 1960). Head-fold stage rat embryonic shields were transplanted into the anterior chamber of the eye of adult rabbits treated with appropriated doses of cortisone in order to suppress immunological reaction. Because the anterior chamber of the eye of the rabbit is much larger than that of the rat, the transplanted rat embryo has more chance to float freely in the aqueous humour than to settle down and become incorporated into the iris. Only those grafts which previously penetrated deeply into the host iris displayed remarkable growth and differentiation into teratome-like structures. On the contrary, freely floating grafts with no contact with the host iris only survived and probably kept growing, but did not undergo histological differentiation (Fig. 4).

In both intraocular homo- and heterografts, incompletely

differentiated embryonic tissues (poorly differentiated epithelia, mesenchyme, immature neural tissue) could be found. Nests of undifferentiated cells, like those occurring in murine teratocarcinomas, were never observed.

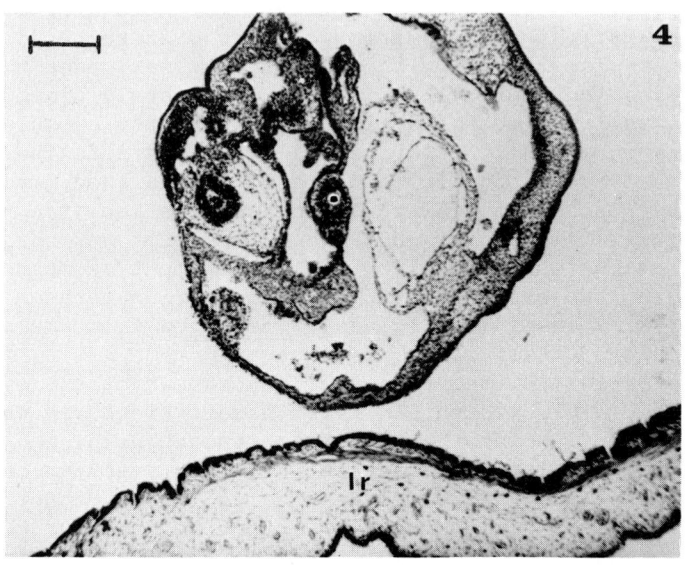

Figure 4. *A head-fold stage rat embryonic shield transplanted into the rabbit anterior chamber of the eye for 15 days. Note the absence of contact with the host iris (Ir) and the poor histological differentiation within the graft. Scale = 0.2 mm.*

One may conclude from all the above described transplantation experiments that the phenotypic expression of transplanted rat embryonic shields depends both on its developmental stage at the time of isolation, and on the extrauterine site to which it has been transferred. Successful grafts result in the development of teratomas, which consist of more or less differentiated normal tissues. The unique case of a tumour under the kidney capsule, consisting of undifferentiated cells only remains to be explained by more detailed experiments using different strains of rats.

IN VITRO CULTURE OF RAT EMBRYONIC SHIELDS

Because in transplantation experiments the nature of

environmental factors is not clearly definable, and cannot be controlled, we have tried to elaborate an *in vitro* method in which the isolated rat embryonic shields are allowed to realize the developmental potentialities which they display in homografts. Previously well developed techniques for the *in vitro* cultivation of pre-implantation mammalian embryos (Brinster, 1970; Biggers et al., 1971; Hsu, 1972) are not applicable at the stage of germ layer formation. After a few unsuccessful attempts with various techniques, we decided to use a slightly modified method of organ culture on a metal grid, as described by Moscona et al. (1965). Isolated embryonic shields (in groups of two to four) were transferred to a piece of lens paper supported by a stainless steel grid, placed in an embryological watch glass. The liquid medium was poured under the grid in amounts sufficient to wet the lens paper. Cultures were incubated for two weeks at 37° in a humid atmosphere composed of 5% CO_2 and 95% air (Skreb and Svajger, 1973).

The explants were cultivated in buffered, chemically defined media (Tyrode's saline, Eagle's minimal essential medium, Ham's F 10 medium) for a limited time span, and then transferred to one of the same above mentioned media enriched with the isologous rat serum. Rat embryonic shields did not survive the initial 24-hour cultivation in Tyrode's saline alone. If this period was reduced to 6 hours, they subsequently developed well in the serum-enriched medium. In Eagle's or Ham's medium the explants could be cultivated for 48 hours without any remarkable negative effect on subsequent differentiation. After having tested different sera in different concentrations, we concluded that the best results could be obtained when at least 20% of the homologous serum was present in the nutrient medium.

Unfortunately, the results were not uniform, even under apparently identical culture conditions. This could not be correlated with the sex and the age of the donors of sera. As far as the histological differentiation in explants is concerned, it was markedly inferior to that in renal homografts, both qualitatively and quantitatively (Fig. 5). After two weeks *in vitro*, the explants formed only small growths consisting of tissues which had often not yet attained their terminal organotypical degree of differentiation. In more than 300 explants of head-fold stage embryonic shields, the following tissues were observed: epidermis (80%), gut and cartilage (60%), neural tissue (55%), skeletal muscle (25%), and in rare cases, gland-like epithelial structures. Heartbeat was noticed in almost every explant after 48 hours. The behaviour of early primitive streak embryonic shields *in vitro* is similar to their behaviour in the anterior chamber of the eye: the resulting small teratomas lacked mesodermal tissue-derivatives.

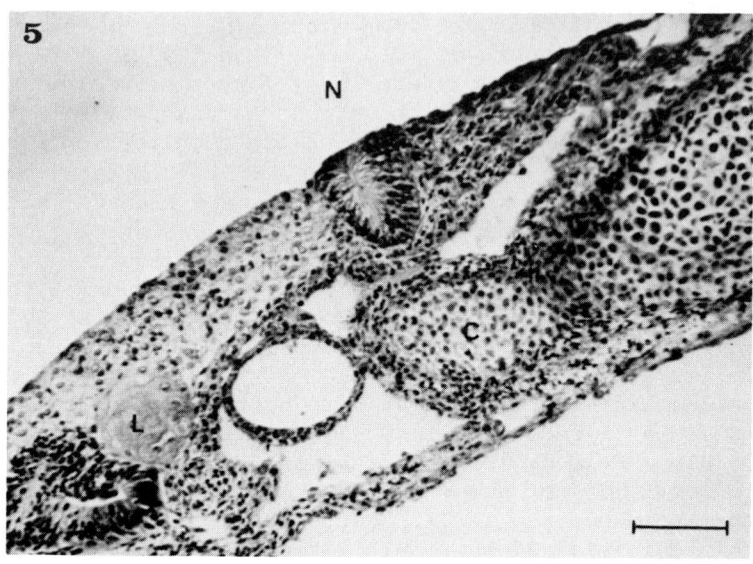

Figure 5. Detail of a tumour derived from a head-fold stage rat embryonic shield explanted for 15 days on lens paper supported by a metal grid (Eagle's minimal essential medium + rat serum, 1 : 1). C, cartilage; L, lentoid; N, immature neural tissue. Scale = 0.1 mm.

If, instead of lying on a metal grid, the explants freely float in the liquid medium, the results strongly resemble those obtained with non-incorporated rat embryonic shields in the rabbit anterior eye chamber: the explants survive and grow, but the histological differentiation is poor (Fig. 6).

In our last experimental series, we transplanted under the kidney capsule embryonic shields which had previously been cultivated *in vitro* for one or two weeks. After cultivation *in vitro* for only a week, the resulting growth and differentiation under the kidney capsule was almost the same as it would have been without the precultivation of grafts. After two weeks of *in vitro* cultivation, histological differentiation, which had already begun, could only be improved to some degree under the kidney capsule. However, the degree of differentiation remained inferior to that observed in renal homografts without precultivation. The re-transplanted two-

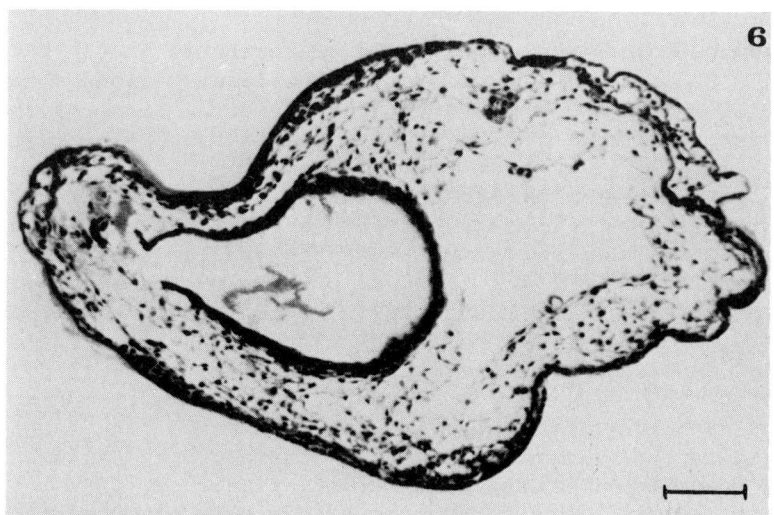

Figure 6. A head-fold stage rat embryonic shield explanted for 15 days in liquid medium. Note the lack of histological differentiation. Scale = 0.1 mm.

layered embryonic shields (pre-primitive streak or early primitive streak stages) could never give rise to mesodermal derivatives. In re-transplants of the three-layered (head-fold stage) embryonic shields, the skeletal muscle did not display any improvement of the differentiation which began *in vitro*. We can therefore speak of a certain stability of such an imperfect differentiation.

CONCLUSIONS

As pointed out in the introduction to this review, the principal aim of the experiments described has been to analyze the differentiation of early rat embryos in extrauterine sites rather than to consider problems of the experimental induction of tumours. We have therefore paid very little attention to the nature of the teratomas obtained, other than their histological composition and organization. Also, we have not tested systematically either the behaviour of selected rat strains or the environmental influences exerted by all available extrauterine sites (testis, omentum). Finally we have not tested the re-transplantability of teratomas obtained.

If, however, we try to answer the questions formulated in the introduction, we may conclude that our work was not in vain, because the results obtained seem to be relevant to the problem of experimental induction of teratomas in the rat.

1) The quality and quantity of differentiation in teratomas is influenced by both the developmental stage of the transferred embryo and the type of the extrauterine environment.

2) An extrauterine environment which would regularly support the growth of rat embryonic cells without their differentiation, was not found. The uncontrolled growth of undifferentiated cells was observed only in one renal homograft. Clusters of undifferentiated cells, which are common in grafts early after transfer, regularly disappear in the course of further development in the extrauterine site. Poorly differentiated tissues found in some experimental conditions (grafts in the rabbit eye or in the *in vitro* explants) cannot be considered as undifferentiated cells, but rather as cells undergoing retarded or abortive differentiation.

3) The isolated rat embryonic shield interacts with the changing extrinsic factors following the rule of 'more or less' differentiation, rather than that of an 'all or none' reaction.

Finally, it must be pointed out that none of the results described definitely excludes the possibility of obtaining re-transplantable teratocarcinomas in rats similar to those which have been produced in mice.

ACKNOWLEDGEMENT

The authors' original work was partly supported by NIH PL 480 Research Agreement Grant No. 02-038-1.

REFERENCES

Biggers, J. D., Whitten, W. K., Whittingham, D. G. (1971). The culture of mouse embryos *in vitro*. In "Methods in Mammalian Embryology" (Ed. J. C. Daniel, Jr.) Freeman & Co., San Francisco, p. 86.

Brinster, R. L. (1970). *In vitro* cultivation of mammalian ova. Advan. Biosciences 4, 199.

Damjanov, I., Solter, D. (1974). Experimental teratoma. Curr. Topics Path. 59, 69.

Fawcett, D. W. (1950). The development of mouse ova under the capsule of the kidney. Anat. Rec. 108, 71.

Grobstein, C. (1951). Intraocular growth and differentiation of the mouse embryonic shield implanted directly and following *in vitro* cultivation. J. Exp. Zool. 116, 501.

Hsu, Y. (1972). Differentiation *in vitro* of mouse embryos beyond the implantation stage. Nature (London) 239, 200.
Kirby, D. R. S. (1962). The influence of the uterine environment on the development of mouse eggs. J. Embryol. Exp. Morph. 10, 416.
Levak-Svajger, B., Skreb, N. (1965). Intraocular differentiation of rat egg cylinders. J. Embryol. Exp. Morph. 13, 243.
Levak-Svajger, B., Svajger, A. (1971). Differentiation of endodermal tissues in homografts of the primitive ectoderm from two-layered rat embryonic shields. Experientia 27, 683.
Levak-Svajger, B., Svajger, A. (1974). Investigation on the origin of the definitive endoderm in the rat embryo. J. Embryol. Exp. Morph. 32, 445.
Levak-Svajger, B., Svajger, A., Skreb, N. (1969). Separation of germ layers in presomite rat embryos. Experientia 25, 1311.
Moscona, A., Trowell, O. A., Willmer, E. N. (1965). Methods. In "Cells and Tissues in Culture" (Ed. E. N. Willmer). Academic Press, New York. Vol. 1, p. 19.
Nicholas, J. S. (1942). Experiments on developing rats. IV. The growth and differentiation of eggs and egg-cylinders when transplanted under the kidney capsule. J. Exp. Zool. 90, 41.
Nicholas, J. S., Rudnick, D. (1933). The development of embryonic rat tissues upon the chick chorioallantoic. J. Exp. Zool. 66, 193.
Skreb, N., Levak, B. (1960). Etude sur la differentiation des jeunes embryons du rat blanc dans l'heterogreffe endoculaire. C. R. Seanc. Soc. Biol. 154, 1750.
Skreb, N., Svajger, A. (1973). Histogenetic capacity of rat and mouse embryonic shields cultivated *in vitro*. Wilhelm Roux'. Archiv 173, 228.
Skreb, N., Svajger, A., Levak-Svajger, B. (1971). Growth and differentiation of rat egg-cylinders under the kidney capsule. J. Embryol. Exp. Morph. 25, 47.
Svajger, A., Levak-Svajger, B. (1974). Regional developmental capacities of the rat embryonic endoderm at the headfold stage. J. Embryol. Exp. Morph. 32, 461.
Usadel, K. H., Rockert, H., Obert, I., Schoffling, K. (1970). Uber die Entwicklung isologer Zellsuspensions-Transplantate aus embryonalen Darm-und Magenanlagen. Z. Ges. Exp. Med. 153, 273.

III.
Surface Antigens On Embryos and Teratomas

EXPRESSION OF A CELL SURFACE ANTIGEN COMMON TO PRIMITIVE MOUSE TERATOCARCINOMA CELLS AND CLEAVAGE EMBRYOS DURING EMBRYOGENESIS AND SPERMATOGENESIS

CHARLES BABINET, HUBERT CONDAMINE
MARC FELLOUS, GABRIEL GACHELIN,
ROLF KEMLER AND FRANCOIS JACOB

Unite de Genetique Cellulaire de l'Institut
Pasteur et du College de France
25 rue du Docteur Roux
75015 Paris, France

INTRODUCTION

Antisera obtained by hyperimmunization of male strain 129 mice with primitive teratocarcinoma cells have allowed the detection of a cell surface antigen (F9 antigen), common to all primitive teratocarcinoma cell lines (PTC) so far tested, to mouse eggs at the morula stage and to mouse sperm (Artzt et al., 1973). This antigen could not be detected on several mouse tumor cell lines, or on differentiated cells derived from teratomas (Artzt et al., 1973). It could thus be considered as an early embryonic antigen, and has been tentatively identified as a direct or indirect product of the $+^{tw32}$ gene of the mouse T-locus (Artzt et al., 1974).

Surprisingly enough, this early embryonic antigen is expressed on two cell types, morulae and spermatozoa. This raises the question of appearance and disappearance of the F9 antigen during embryogenesis. We report here the results of the use of the anti-PTC serum in the study of the developmental expression of the F9 antigen.

MATERIAL AND METHODS

Animals

129/Sv C.P. is a subline of the original 129 strain. The teratoma OTT 6050, from which all the PTC cell lines used in our laboratory originate, has been induced in this strain and is syngeneic to it (Stevens, 1970).

Sl^d/Sl^d : Steel-Dickie (Sl^d) character is on the inbred background C3H/He. Homozygotes are sterile due to the absence of primordial germ cells.

Antisera

Anti-PTC serum was produced by hyperimmunization of 129 male mice with irradiated PTC (F9-41) cells. Its properties have been reported elsewhere (Artzt et al., 1973). Negative control sera were either anti-PTC serum extensively absorbed on F9-41 cells or normal mouse serum. All sera were heat inactivated. Rabbit anti-mouse Ig, labelled with fluorescein and goat anti-mouse Ig antibodies labelled with peroxidase are products of Institut Pasteur Production (Paris). These sera were used after absorption on PTC cells, testicular cells and pooled uterine and embryonic cells. Rabbit serum of an animal with a deficiency of C6 was a generous gift of Pr. K. Rother (Institute of Immunology, Univ. of Heidelberg, W. Germany). Sheep anti-rabbit C3 was purchased from Nordic Labs. Both sera were absorbed on F9 cells and on an acetone powder of mouse liver.

Immunostaining of Mouse Embryos

Unfertilized eggs and fertilized preimplantation embryos were obtained from 129 mice and processed, prior to labelling, as described elsewhere (Artzt et al., 1973). The F9 antigen was studied on these samples by using conventional immunological methods (indirect staining with peroxidase or fluorescein). When these techniques failed to demonstrate specific labelling, a more sensitive method based on C3 bind-

---

ABBREVIATIONS AND TERMINOLOGY

PTC - primitive teratocarcinoma cell lines.

The day of observation of the sperm plug is the first day of pregnancy.

ing to surface bound antibodies was used (Klein et al., 1974). Dilutions of antiserum were 1:30 to 1:3200.

Frozen sections of blastocysts and of postimplantation embryos were made in cryoform using an IEC cryotome. Frozen sections were examined for the F9 antigen by indirect immunofluorescence or indirect immunoperoxidase techniques. Dilutions of antiserum were 1:50 to 1:150.

Immunostaining of Frozen Sections of Mouse Testes

Frozen sections of mouse testes (10-12 µm) were air dried and immediately examined for F9 antigen by the same immunofluorescence methods. Dilutions of antiserum were 1:40 to 1:320.

Absorption of Antisera on Testicular Cells

Whenever possible, testicular cells, isolated according to Artzt et al., (1973), were used in massive and quantitative absorption of anti-PTC serum. As a control of specificity of the absorption, an anti-$H-2^k$ serum (129 strain is $H-2^b$) was also absorbed on the same cells. Absorptions were made on ice for one hour.

Residual cytotoxic activities (anti-PTC and anti $H-2^k$) were determined in a microcytotoxicity test as described previously (Artzt et al., 1973), on the proper target cells (F9 and C3H lymph node lymphocytes, respectively).

RESULTS

Expression of F9 Antigen on Early Embryos

Unfertilized eggs. We have previously reported that the F9 antigen was not detectable on unfertilized eggs by immunoperoxidase assay (Artzt et al., 1973). We also failed to detect it by using immunofluorescence and the more sensitive indirect anti-C3 test. In the latter test, the lowest dilution of sera used was 1:30. It is not ruled out that traces of antigen could be detected with higher concentrations of anti-PTC serum.

Fertilized one cell eggs. We failed to detect the presence of F9 antigen on these samples by using the same methods and techniques.

Two-cell eggs and cleavage embryos. By contrast, two-cell eggs were positive in all types of tests performed. The F9 antigen is expressed at their surface and can be detected at a dilution of 1:320 in the anti-C3 test. This finding indicates that synthesis or integration into the membrane of F9 antigen takes place by this stage.

Morulae. The amount of F9 antigen present on the surface of these cells increases very quickly and tends to a maximum at the late morula stage (16 cells). Specific immunostaining is observed with an antiserum diluted 1:3200 (Artzt et al., 1973).

Blastocysts. Early and late blastocysts have been examined in several ways. Intact blastocysts were specifically labelled by anti-PTC serum in the same range of dilutions as morulae (down to 1:3200). Frozen sections of blastocysts were examined after immunostaining with fluorescein-labelled or peroxidase-labelled antibodies. Both trophectoderm and inner cell mass cells were equally stained. The labelling was confined to the membranes.

Implanted embryos. As soon as the embryo has implanted (5th day), expression of the F9 antigen is made difficult to study. Since F9 activity is not retained by fixed cells, frozen sections have to be used, and the structures are not well preserved. Also, decidua and embryonic membranes were found to be heavily fluorescent with both immune and preimmune sera. The embryo proper is not labelled with anti-PTC serum at the earliest stage after implantation studied (6th day). But, for the reasons given above, the meaning of this result is questionable. At most, it can be suggested that the antigen is no longer expressed on the majority of the embryo cells and that its presence on parietal yolk sac cells and trophoblast cells is not ruled out.

Expression of F9 Antigen on Testicular Cells

At the other end of embryonic development, Artzt et al., (1974) have reported that the F9 antigen was expressed on spermatozoa. It is exclusively located on the post acrosomal zone (Fellous et al., 1974). However, its time of synthesis was not determined. Again, examination of frozen sections after immunostaining is made difficult by a high background and the secretion by spermatids of autoantigens that bind to autoantibodies. The slow maturation of germ cells in postnatal testes allows their study at ages when autoantigens are not yet synthesized and germ cells are blocked at essentially the same stage in spermatogenesis.

Newborn animals. In newborn animals, the testis contains rare large gonocytes and peripheral supporting cells. In frozen sections, these large cells showed a very faint, but specific, labelling with anti-PTC serum while supporting cells were not stained (dilution 1:40). Testicular cell suspensions from newborn animals (300,000 gonocytes) completely absorbed the anti-PTC activity of the anti-serum (60 μl, dilution 1:600). The same suspension failed to absorb anti-$H-2^k$ activity.

6 day animals. In these animals, all gonocytes have disappeared, and their descendants, spermatogonia, are actively proliferating. On immunostained frozen sections, spermatogonia were all intensely stained on the membrane. Supporting cells were not. Absorption experiments were performed on suspensions of cells (mainly spermatogonia, since most Sertoli cells are lyzed). Again, all of the activity was removed.

10-12 day animals. At this age, 25 to 40% of the seminiferous tubules contain primary spermatocytes at the leptotene stage. Frozen sections stained with anti-PTC serum display a very bright fluorescence, confined to spermatogonia and spermatocytes. The same cells do not exhibit any labelling when treated with the same serum previously absorbed with F9 cells. A quantitative absorption of anti-PTC serum was performed on a suspension of spermatogonia and spermatocytes. About 200,000 cells removed half of the activity, under the conditions used to type PTC cells for F9 antigen (Dexter et al., 1974). This value is close to that obtained with several PTC lines.

Adult animals. These results were confirmed on frozen sections made from normal adult testes, removed from a fertile male. All diploid precursors of spermatozoa were stained by anti PTC serum, as were secondary spermatocytes and spermatids.

Mutant mice lacking germinal line. In order to rule out the possibility of an expression of the F9 antigen on supporting (Sertoli) cells, we studied testes of animals lacking germinal cells, but having a normal development of seminiferous tubules. Testes of a 15-day old animal (Sl^d/Sl^d) were compared to testes of the heterozygote $+/Sl^d$ (C3H) and of the wild type (C3H). The F9 antigen could be easily identified in sections of the two latter animals. By contrast, it could not be detected in the homozygote mutants: the supporting cells were not labelled, indicating that the amount of antigen expressed at their surface is beyond the limits of detection of the methods used.

DISCUSSION

Developmental expression of the F9 antigen could thus

be summarized as follows: the unfertilized egg has no F9 antigen detectable at its surface. After fertilization by a spermatozoon which brings traces of F9 antigen (which are likely to be incorporated in the egg membrane (Yanagimachi et al., 1973), a progressive appearance (synthesis or unmasking) of the cell surface antigen is detected with the first cell division. The antigen is then expressed on all cells, even after they differentiate into trophoblast and inner cell mass. Soon after implantation, the antigen seems to disappear from the surface of most of the cells of the embryo proper.

On male germ cells, the F9 antigen is expressed not only on spermatozoa, but on all diploid precursors and also on gonocytes. This finding suggests that, at least in the male, the F9 antigen is continuously expressed on germ cells, and thus on primordial gonocytes.

The female germ line has not yet been investigated. The presence of traces of F9 antigen on the unfertilized egg cannot be ruled out. If this antigen were to be found on all primordial gonocytes (male and female), the germinal line would be characterized by the continuous expression of the PTC antigen at its surface. On the other hand, none of the adult transplantation antigens is expressed on these same cells. It is thus tempting to hypothesize that the expression of the F9 embryonic antigen and the abs**ence** of adult H-2 antigens provide a way for the embryo to sort out germinal cells from other differentiating tissues.

ACKNOWLEDGEMENTS

This work was supported by grants from the "Centre National de la Recherche Scientifique", the "Delegation Generale a la Recherche Scientifique et Technique" the National Institute of Health and the Andre Meyer Foundation. One of us (Rolf Kemler) is a fellow of the "Deutsche Forschungsgemeinschaft".

REFERENCES

Artzt, K., Bennett, D., Jacob, F. (1974). Primitive teratocarcinoma cells express a differentiation antigen specified by a gene at the T-locus in the mouse. Proc. Nat. Acad. Sci. U.S. 71, 811.

Artzt, K., Dubois, P., Bennett, D., Condamine, H., Babinet, C., Jacob, F. (1973). Surface antigens common to mouse cleavage embryos and primitive teratocarcinoma cells in culture. Proc. Nat. Acad. Sci. U.S. 70, 2988.

Dexter, D. L., Buc-Caron, M. H., Jakob, H., Nicolas, J., Gachelin G. (1974). Teratocarcinome de la souris: etude des antigenes de la surface des cellules a potentialites multiples. Ann. Microbiol. (Inst. Pasteur) 125B, 347.

Fellous, M., Gachelin, G., Buc-Caron, M. H., Dubois, P., Jacob, F. (1974). Similar location of an early embryonic antigen on mouse and human spermatozoa. Develop. Biol., 41, 331.

Klein, G., Giovanella, B. C., Lindahl, T., Malkow, P. J., Singh, S., Stehlin, J. S. (1974). Direct evidence for the presence of Epstein-Barr virus DNA and nuclear antigen in malignant epithelial cells from patients with poorly differentiated carcinoma of the nasopharynx. Proc. Nat. Acad. Sci. U.S. 71, 4737.

Stevens, L. C. (1970). The development of transplantable teratocarcinomas from intratesticular grafts of pre- and post-implantation mouse embryos. Develop. Biol. 21, 364.

Yanagimachi, R., Nicolson, G. L., Noda, Y. D., Fujimoto, M. (1973). Electron microscopic observations of acidic anionic residues on hamster spermatozoa and eggs, before and during fertilization. J. Ultrastruc. Res. 43, 344.

TERATOMA-DEFINED AND TRANSPLANTATION ANTIGENS IN EARLY MOUSE EMBRYOS[1]

MICHAEL EDIDIN AND LINDA R. GOODING[2]

Department of Biology
The Johns Hopkins University
Charles and 34th Streets
Baltimore, Maryland   21218

The mammalian fetus develops to term where immunologic rejection would overtake any other tissue graft. The fetus appears to be protected from rejection by a number of mechanisms, including: immunological modification of the mother, buffering of fetus and mother by an antigenically inert barrier (e.g., the trophoblast), and, at preimplantation stages of development, a degree of antigenic immaturity of embryonic cell surfaces. We have used cells derived from the spontaneous strain 129 teratocarcinoma 402 AX as a sample of the cell surface in early normal development. The response to these cells in transplantation experiments, as well as the reaction with normal embryos of antisera against the tumor, give us some hints as to the degree of antigenic maturation in regard to transplantation antigens that occurs in the first 1/3 of normal development.

In this paper we summarize briefly what is known about the transplantation antigens of early mouse embryos, as well as our previous studies with antigens of 402 AX. These summaries are followed by a demonstration that teratoma-defined antigens are present in early normal embryos, and that they are confined to tissues that later will express strong transplantation antigens.

---

[1]Dedicated to the memory of Joy Palm.
[2]Present address: Division of Immunology, Duke Medical Center, Durham, North Carolina   27710.

MICHAEL EDIDIN AND LINDA GOODING

THE STRENGTH AND VARIABILITY OF TRANSPLANTATION ANTIGENS

Rejection of grafts between individuals of a single species has been clearly shown to be due to an immune response by the recipient against antigens of the graft (Medawar, 1958). The genetics of these tranplantation antigens is best understood for mice. From studies of inbred animals, it appears that over 60 genetic loci affect transplantation alloantigens. However, one of these loci, the H-2 complex, is paramount in determining graft survival. Compatibility for H-2 ensures prolonged survival of grafts in many strain combinations, while incompatibility only at H-2 invariably leads to rapid graft rejection (Klein, 1975a). Indeed, careful studies indicate that a mismatch at four non-H-2 loci produces no faster rejection of skin allografts than a single H-2 mismatch (Graff et al., 1966). A second striking aspect of H-2 is its high degree of polymorphism and of genetic complexity. The complex consists of at least 7 different genetic regions. Products of some of these regions have classically been detected with 'anti H-2' antisera, while products of other regions appear to affect only cellular immune responses, or are detectable only with antisera prepared in special strain combinations (Shreffler and David, 1975). The permutation of alleles of the seven known regions of H-2 leads to a large number of possible combinations of haplotypes. More than 25 are known for inbred strains only, and there appear to be still more present in wild mice (Klein, 1970). In contrast, other H-loci have few alleles or pseudo-alleles.

Homologues of H-2 are found in all other mammals thus far studied, notably in humans, where over 300 HL-A haplotypes have been catalogued, and in rats, dogs, monkeys (Snell, 1968). Though the biological function of the strong antigens determined by H-2 and its homologues is unclear, it is evident that the chance of compatibility between a fetus and its mother in random-breeding populations is small, and that the mammalian conceptus usually develops despite the genetic potential for graft rejection, directed against strong transplantation antigens, and effected by both immune lymphocytes and by antibody. Though the fetus is protected in later development by specialized membranes, and perhaps by modifica-

---

ABBREVIATIONS AND TERMINOLOGY

ICM - inner cell mass of the blastocyst.
TMR - tetramethylrhodamine.

tion of maternal immune responses, it is potentially vulnerable to rejection in early development. At least part of the mechanism evolved to prevent this appears to be delay in expression of strong transplantation antigens until after implantation of the conceptus; furthermore, H-2 expression is restricted to derivatives of the inner cell mass, which are surrounded by less antigenic extraembryonic cells.

TRANSPLANTATION ANTIGENS IN EARLY EMBRYOS

Transplantation antigens have been assayed in embryos by two approaches, which may well detect products of different genetic regions. Ectopic transplantation of stages up to blastocyst of one inbred strain to recipients of another results in growth of a benign teratoma, followed by its eventual rejection. This indicates at least that tranplantation antigens developed after making the graft. However, if grafts are made to allogeneic recipients previously immunized to adult cells of the graft donor strain, then rejection can be detected within a few days of grafting, and no differentiation or proliferation of tissue occurs (Edidin, 1964a; Kirby et al., 1966; Simmons and Russell, 1966). Hence, some transplantation antigens must be present on the conceptus from zygote stage on. On the other hand, if grafts are made between strain combinations that differ only at H-2, presensitized or not, rejection is delayed for several days after grafting blastocysts (Searle et al., 1974) or egg cylinder stages (Patthey and Edidin, 1973). It appears that H-2 antigens are absent from embryos through 6 days of development, and that they are expressed in quantities sufficient to trigger immediate graft rejection in presensitized hosts by day 7 after conception, or around the time of somite formation. Rejection of embryo grafts across multiple transplantation antigen barriers then must be due to incompatibilities at non-H-2 loci, and, indeed, grafts matched only for H-2 are rapidly rejected by their preimmunized hosts (Searle et al., 1974).

Tissue transplantation experiments are paralleled by serologic measurements of antigens in early embryos. Palm and co-workers (1971), using one of the few antisera reacting with non-H-2 antigens, unequivocally demonstrated the presence of H-3 and H-6 antigens on zygotes and cleavage stage embryos. On the other hand, Heyner (1973) did not find serologically detectable H-2 antigens on cultured blastocysts until they had developed to the egg-cylinder stage, equivalent to about 6 days of normal pregnancy. All results with antisera must be taken with caution. Many antisera were made on the assumption that only H-2 antigens provoke circulating antibodies; this assumption is not strictly true (Palm et al., 1971). Furthermore, it is not clear that even antibodies directed only to H-2 complex products react with the same cell surface molecules. Products of the D, K and I regions may all react with alloantisera, and as a result no adequate study has yet been done of specific H-2 antigens in early development (see Edidin et al., 1974, and the remarks by Askonas in the discussion). Still, the results with antisera roughly parallel those obtained by grafting. They imply that by any criterion preimplantation embryos are H-2 null. Aside from normal embryos and teratomas (Artzt and Jacob, 1974; Edidin et al., 1974) no other cells are known to be totally lacking in H-2 (Edidin, 1972; Klein, 1975b).

ARE THERE H-2 ANTIGENS ON TERATOMA CELLS?

As hinted at the end of the previous section, the answer to the question of this section is a barely qualified 'no'. Two strain 129 teratomas have been tested for H-2 using antibody and grafting techniques. Anti-H-2 serum and complement are no more effective than normal mouse serum and complement in lysing F9 cells, which are derived from teratoma OTT 6050. More significantly, these cells do not absorb out anti-H-2 activity in serum, even when ratios of 1:2 cells:diluted serum are used (Artzt and Jacob, 1974). A positive control of poorly antigenic cells removed more than half of the antibody activity of the test serum when used for absorption.

We have examined cells of both 402 AX and OTT 6050 for H-2, using indirect immunofluorescence and grafting techniques. Some 402 AX cells, in a cultured subline of the tumor that still differentiates morphologically (Searles and Edidin, unpublished), bind anti-H-2 antibody but only around 8% of cells in any population examined have been H-2 positive by fluorescence. Similar populations may be tested *in vivo* for H-2 complex gene products. Here we follow a classic protocol for presensitization to skin grafts. The putative antigen is given 3 days before grafting skin. Its effect is detected in the rate at which grafts are rejected, compared to the rate for grafts on control, untreated animals. One further refinement is necessary to test for H-2 only; grafts are made between members of congenic-resistant strains (Snell and Bunker, 1965) bred to differ only at the H-2 region. Figure 1 shows the survival rates for skin grafts made from B10 to to B10.Br animals. B10 animals share the H-2^b haplotype with strain 129. Graft recipients preimmunized with 4×10^4 129 lymphocytes (cells rich in H-2) reject their grafts at an accelerated rate, compared to controls. Animals receiving 6×10^6 teratoma cells appear to reject grafts at approximately the same rate. Hence, on a population level, teratoma cells seem to be about 300-fold less antigenic than lymphocytes. This level of H-2 antigenicity could be found either if each cell bore about 100 H-2 molecules on its surface, or if most cells were H-2 null and a few had H-2 levels approaching those found for its human homologue, HL-A, on lymphocytes, probably around 50,000/cell (Sanderson, 1974).

A similar study has been done of cells dissociated from embryoid bodies of OTT 6050. The results presented in Figure 2 suggest that, on the average, these cells are even less H-2-active than 402 AX cells. Indeed, though the embryoid bodies were cultured overnight, in order to remove host (129) macrophages adherent to their surface, all of the activity found in this population could be due to a few passenger cells. The results are in good agreement with those of Artzt and Jacob (1974), on F9, though, of course, a direct grafting test for H-2 on F9 would strengthen the comparison.

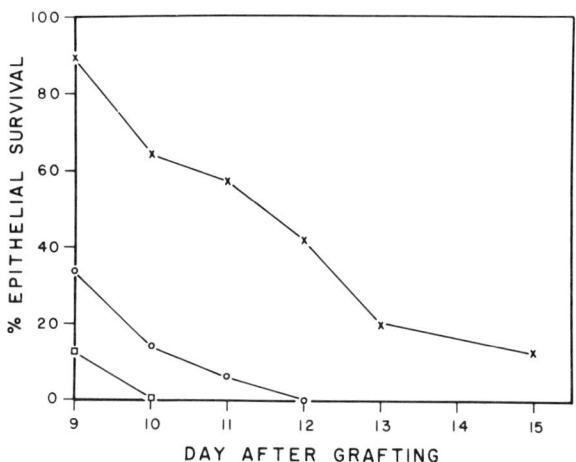

Figure 1. Rejection of B10 skin (H-2^b) by B10.Br (H-2^k) animals. Recipients were either untreated, x , or injected three days prior to grafting with □ , 4×10^4 129 (H-2^b) lymphocytes, or O , 6×10^6 402 AX teratoma cells. Survival of graft epithelia was estimated by eye daily after the bandages were removed at 9 days after grafting.

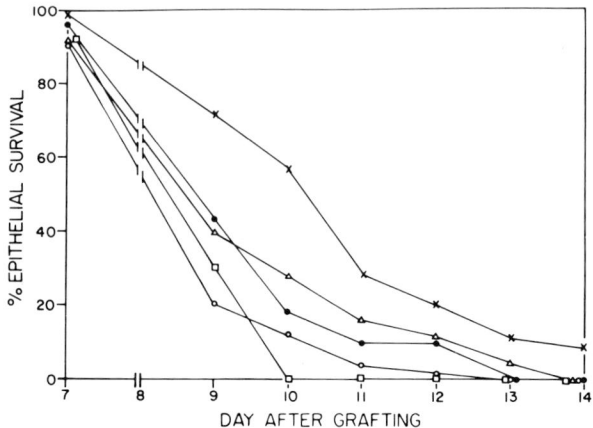

Figure 2. Rejection of B10 skin by B10.Br recipients. Animals were either untreated, x, or injected three days prior to grafting with △, 2×10^4 lymphocytes, □, 4×10^4 lymphocytes, ○, 6×10^6 402 AX cells, ●, 6×10^6 OTT 6050 cells. Survival of graft epithelia was estimated daily following removal of bandages 6 days after grafting.

The near absence of H-2 in teratoma cell populations is all the more striking when contrasted with the persistence of these antigens on tumor and cultured cells exposed to direct selection against their antigens. No H-2 homozygous cell has ever been prepared that lacks these antigens (Klein, 1975b). The only other cells that appear to lack H-2 and homologous antigens are those of the early embryo mentioned previously and those of the trophoblast (Jenkinson and Billington, 1974).

RELATIONSHIP BETWEEN TERATOMA-DEFINED ANTIGENS AND H-2

We have prepared a hetero-antiserum against tumor 402 AX cells by immunizing rabbits with live tumor cells and a bacterial adjuvant (Gooding and Edidin, 1974). The serum obtained reacts with at least three different antigens, which may be dissected by absorption on a set of tumors of various origins (Table 1). Of course each of these 'antigens', may in turn be a family of substances, reacting with a corresponding family of antibodies. However, at present, absorptions have not split the serum into further subsets, and so

TABLE 1

DISTRIBUTION OF TERATOMA-DEFINED ANTIGENS
ON A SERIES OF TUMOR CELLS

Cells[a]	Antigens Present	% Cells Reacting with Antiserum Absorbed on[b]			
		-	Cl 1d	Hepatoma	Teratoma
Cl 1d	I	97	0	0	0
Melanoma B16	I	86	8	4	3
Hepatoma BW7756	I,II	100	99	6	10
Teratoma 402 AX	I,II,III	100	93	68	0

[a]For details of tumors used in these experiments see Gooding and Edidin (1974).

[b]Cells were treated with 1/20 antiserum and fluorescent goat-anti-mouse IgG.

for convenience we use the singular form, e.g. antigen I. One of these antigens, antigen I, is present on the surface of several transformed cell lines; in one of these lines, Cl 1d (Dubbs and Kit, 1964), antigen I is further exposed by trypsin, and co-caps with H-2 antigens. Here there seems to be a physical association between the tumor antigen and H-2, leading to co-migration when one of the two partners is aggregated by specific antibody. Also, the tumor antigen is less exposed at the surface than is its associated strong histocompatibility antigen. Thus, antigen I appears to be attached to or associated with H-2 so that the former lies deeper in the membrane than the latter. We speculate that this physical association reflects an earlier step in the insertion of newly-synthesized H-2 antigens into the cell surface. Perhaps H-2 cannot be inserted directly into a protein-free region of the lipid bilayer, but rather, must first complex with antigen I in order to be delivered to the surface in correct orientation.

Study of antigen I in embryos reinforces our bias that the antigen forms some sort of precursor to H-2, either because H-2 must insert in the membrane at the site of an antigen I molecule, or because antigen I functions in a way equivalent to H-2 in early stages of development. (Note that the former interpretation begs the question of H-2 function at the cell surface; the latter interpretation asserts that H-2 and its homologues have a function, but do not define this function).

TERATOMA-DEFINED ANTIGENS ON EARLY MOUSE EMBRYOS

Three preparations of antiserum have been used for our studies. One is the whole antiserum to teratoma, reacting with all three antigens. A second is prepared from the first by absorption with Cl ld. This serum reacts with antigens II and III but not antigen I. Finally, since all positive tumors tested bear antigen I, we prepared an anti-I reagent by absorbing whole serum to a Sepharose-coupled extract (100,000 g x 120 min supernatant) of melanoma cells, which contains only antigen I (Gooding and Edidin, 1974). Anti-I antibody binding to the affinity column could be eluted in 1N HCl, after washing the column thorough with 0.1 M sodium acetate pH 3.6 and 0.1M acetic acid to remove unbound and weakly bound proteins. The eluted specific antibody was completely absorbed by Cl ld as tested against these cells, or against teratoma 402 AX cells.

Ova and early embryos through blastocyst stage were obtained from both random-bred CF1 mice and from (HTG/HTI) F1 mice. Inner cell mass (ICM) was examined in 'post-implantation' embryos cultured on collagen for five days, resulting in migration of their trophoblast layer off the inner cell mass and onto the collagen (Hsu, 1972). Binding of antiserum to cells at all stages was detected with fluorescein- or tetramethylrhodamine (TMR)-conjugated goat anti-rabbit IgG prepared in this laboratory.

The reactions of whole anti-teratoma serum and of purified anti-I antibody are summarized in Table 2. Representative examples of positive reactions with unfertilized eggs, morulae, blastocysts and post-implantation inner cell mass are shown in Figures 3-8. Control reactions, using teratoma absorbed anti-teratoma serum or Cl ld absorbed anti-I, were

Figure 3. Unfertilized ovum stained with anti-I antibody and TMR-conjugated anti-globulin.
Figure 4. A four-cell embryo stained with anti-I antibody and fluorescein conjugated anti-globulin.
Figure 5. A morula stained as in Figure 4.
Figure 6. Blastocyst stained with whole anti-teratoma antiserum and TMR-conjugated anti-globulin.
Figure 7. A 'post-implantation' embryo. A blastocyst was cultured for 5 days on collagen to obtain this stage. Trophoblast cells have migrated into the collagen, exposing the inner cell mass.
Figure 8. The cultured blastocyst of Figure 7 stained with anti-teratoma antiserum. Trophoblast cells do not react.

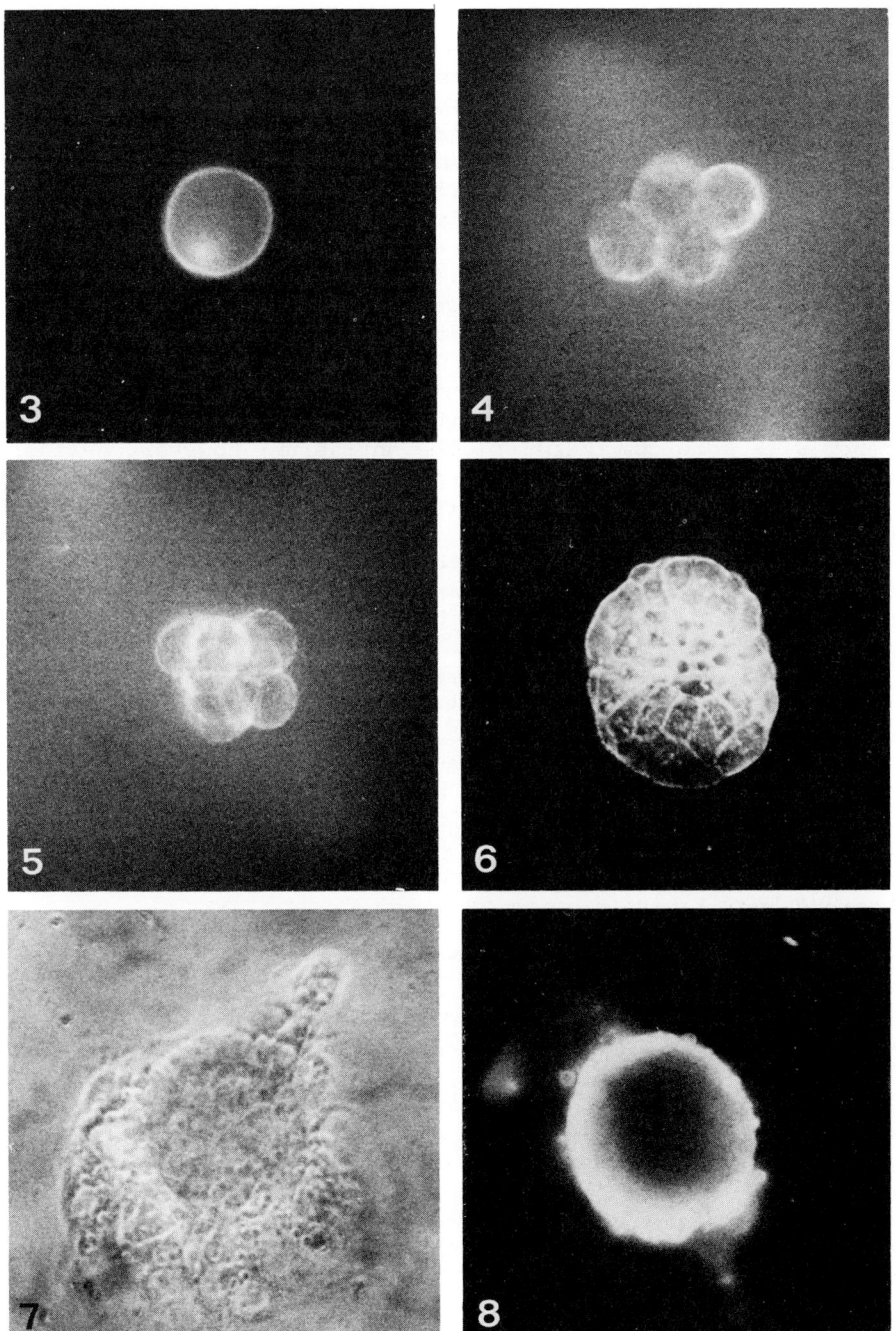

effectively blank; no difficulty was experienced with background or 'non-specific' fluorescence.

The data of Table 2 are readily summarized with respect to antigen I. The antigen is present on unfertilized eggs, blastomeres of cleavage stages and inner cell mass cells of post-implantation embryos. Antigen I is absent from trophoblast giant cells and from trophectoderm cells of late, hatched, blastocysts. Trophectoderm cells of early blastocysts have not been completely analyzed; it seems that trophectoderm overlying the inner cell mass bears antigen I, and no other teratoma-defined antigens, while trophectoderm not in contact with the inner cell mass contains antigen II, and could bear antigen I as well.

We have also examined later embryos with whole anti-teratoma serum. Cells of embryos up to 8 days of development react with some components of the antiserum; however, neither Cl 1d absorbed, nor antigen-I specific sera have yet been tested against cells of this stage. In any case, it appears

TABLE 2

REACTION OF ANTI-TERATOMA ANTISERA
WITH EARLY MOUSE EMBRYO CELLS[a]

Stage	Whole Antiserum Absorbed on:			Anti-antigen I Antibody	
	−	Cl 1d	Hepatoma	Unabsorbed	Cl 1d Absorbed
Unfertilized Ovum	+	±	−	+	−
Cleavage (4-16 cells)	+	−	−	+	−
Early unhatched blastocyst:					
Trophectoderm cells adjacent to ICM	+	−	−	nd[b]	nd
Trophectoderm cells not in contact with ICM	+	+	−	nd[b]	nd
Hatched blastocyst:					
Trophectoderm cells	+	+	−	−	−
Post-implantation ICM	+	+	−	+	−
Trophoblast giant cells	−	−	−	−	−

[a]Embryos were reacted with 1/10 antiserum or equivalent amount of anti-I antibody as titered on Cl 1d cells.

[b]nd = not determined.

that all reactivity of mouse embryos with anti-402 AX is lost by 9 days of development, whether embryo whole mounts or dissociated cells are examined.

SPECULATION: THE RELATIONSHIP BETWEEN TERATOMA ANTIGEN I AND H-2 IN NORMAL DEVELOPMENT

The data on appearance of antigen I in development add some weight to our idea that this antigen appears in cell membranes as a precursor to H-2. The antigen is present at all stages that are shown to lack H-2, and is lost, at the latest, at a stage which provokes *in vivo* cellular responses and in which low levels of H-2 are easily detectable by absorption (Edidin, 1964b). However, we present here only the natural history of a surface antigen. Its relationship to H-2 is suggested by this description but must be tested directly in a cell system that can be manipulated to shift from H-2 null, antigen I positive to H-2 positive, antigen I negative. The embryo itself is formidable material for this experiment; in it, a variety of cell types proceeds ansynchronously through many differentiation steps. Rather, we look to the teratocarcinoma model system in terms of its potential for differentiation and differential expression of surface antigens. Having used teratoma 402 AX to define a surface antigen of early embryo cells, we now hope to use it further in mapping the fate of that antigen in development.

ACKNOWLEDGEMENTS

We thank Ms. Lisa Schwender for skin grafting our animals. Original work reported here was supported by NIH grant AM11202 to ME, and a training grant to the Department of Biology. This is contribution no. 831 from the Department of Biology.

REFERENCES

Artzt, K., Jacob, F. (1974). Absence of serologically detectable H-2 on primitive teratocarcinoma cells in culture. Transplantation 17, 632.
Dubbs, D. R., Kit, S. (19 4). Effect of halogenated pyrimidines and thymidine on growth of L-cells and a subline lacking thymidine kinase. Exp. Cell Res. 33, 19.
Edidin, M. (1964a). Transplantation antigens in the mouse

embryo. J. Embryol. Exp. Morph. 12, 309.

Edidin, M. (1964b). Transplantation antigen levels in the early mouse embryo. Transplantation 2, 627.

Edidin, M. (1972). The tissue distribution and cellular location of transplantation antigens. In "Transplantation Antigens" (Eds. B. D. Kahan and R. Reisfeld). Academic Press, New York, p. 125.

Edidin, M., Gooding, L. R., Johnson, M. H. (1974). Surface antigens of normal early embryos and a tumor model system useful for their further study. Acta Endocrin. 78, Suppl. 336.

Graff, R., Silvers, W. K., Billingham, R. E., Hildemann, W. H., Snell, G. D. (1966). The cumulative effect of histocompatibility antigens. Transplantation 4, 605.

Gooding, L. R., Edidin, M. (1974). Cell surface antigens of a mouse testicular teratoma. Identification of an antigen physically associated with H-2 antigens on tumor cells. J. Exp. Med. 140, 61.

Heyner, S. (1973). Detection of H-2 antigens on the cells of the early mouse embryo. Transplantation 16, 675.

Hsu, Y.-C. (1973). Differentation *in vitro* of mouse embryos to the stage of early somite. Devel. Biol. 33, 403.

Jenkinson, E. J., Billington, W. D. (1974). Differential susceptibility of mouse trophoblast and embryonic tissue to immune cell lysis. Transplantation 18, 286.

Kirby, D. R. S., Billington, W. D., James, D. A. (1966). Transplantation of eggs to the kidney and uterus of immunized mice. Transplantation 4, 713.

Klein, J. (1970). Histocompatibility-2 (H-2) polymorphism in wild mice. Science 168, 1362.

Klein, J. (1975a). "Biology of the Mouse Histocompatibility-2 complex". Springer Verlag. New York. Chapter 8.

Klein, J. (1975b). "Biology of the Mouse Histocompatibility-2 complex". Spring Verlag. New York. Chapter 13.

Medawar, P. B. (1958). The immunology of transplantation. The Harvey Lectures 52, 144.

Palm, J., Heyner, S., Brinster, R. L. (1971). Differential immunofluorescence of fertilized mouse eggs with H-2 and non-H-2 antibody. J. Exp. Med. 133, 1282.

Patthey, H., Edidin, M. (1973). Evidence for the time of appearance of H-2 antigens in mouse development. Transplantation 15, 211.

Sanderson, A. R., Welsh, K. L. (1974). Properties of histocompatibility (HL-A) determinants. Transplantation 17, 281.

Searle, R. F., Johnson, M. N., Billington, W. D. Elson, J., Clutterbuck-Jackson, S. (1974). Investigation of H-2 and non-H-2 antigens on the mouse blastocyst. Transplantation 18, 136.

Shreffler, D. C., David, C. S. (1975). The H-2 major histocompatibility complex and the I immune response region. Advan. Immunol. 20, 125.

Simmons, R., Russell, P. S. (1966). The histocompatibility antigens of fertilized mouse eggs and trophoblast. Ann. N. Y. Acad. Sci. 129, 35.

Snell, G. D., Bunker, H. P. (1965). Histocompatibility genes of mice. V. Five new histocompatibility loci identified by congenic resistant lines of a C57BL/10 background. Transplantation 3, 235.

Snell, G. D. (1968). The H-2 locus of the mouse: Observations and speculations concerning its comparative genetics and its polymorphism. Folia Biologia (Praha) 14, 335.

SURFACE ANTIGENS OF BLASTOCYST-DERIVED CELL LINES

RICHARD A. MILLER[1], FRANK H. RUDDLE[1] AND
MICHAEL I. SHERMAN[2]

[1]Department of Human Genetics, Yale University,
New Haven, Ct.   06520
and [2]Roche Institute of Molecular Biology,
Nutley, N. J.   07110

The recent availability of cultured cell lines derived from early mouse embryos (Sherman, 1975a,b) and teratocarcinomas (Rosenthal et al., 1970; Kahan and Ephrussi, 1970) is stimulating a number of new *in vitro* approaches to fundamental developmental problems. Convincing assignment of these cell lines to particular embryonic tissues may well prove difficult, but it is a necessary first step towards these developmental studies to show that such analogies can usefully be made. Phenotypic characterization of these lines will also provide the epigenetic markers needed to study developmental transitions *in vitro*.

Artzt and her colleagues (1973) have demonstrated that one such cell line, F9, an embryonal carcinoma derived ultimately from a six-day embryo, bears a surface antigen not found on teratoma-derived cell lines composed of differentiated cells, but which is present on mouse morulae. The key immunologic tactic in their work was the injection of F9 cells into *syngeneic* (129/Sv male) mice, in order to restrict the resulting immune response to those antigens not encountered by the animal during the development of immune self-tolerance. These antigens might include determinants found on cells in immunologically privileged sites in the adult (e.g. spermatozoa) or on embryonic cells which are no longer represented among adult somatic tissues. Their subsequent work (Artzt et al., 1974) showed that the expression of the F9 antigen depended on the presence of the wild-type allele of a known recessive lethal, t^{12}, within the complex Brachyury locus (see Gluecksohn-Waelsch and Erickson, 1970).

Since t^{12} homozygous embryos cannot pass from the morula to the blastocyst stage, the F9 ($+t^{12}$) antigen may well play a role in this developmental event.

Sherman (1975a,b) has recently derived continuous cell cultures of indefinite lifespan from 4th day mouse blastocysts. Several of these cell lines can be distinguished from one another by enzymatic, histologic, and immunologic criteria. Figure 1 depicts the culture morphology of two of these lines, MB2 and MB4, and also shows the histological appearance of their malignant variants, MB2T1 and MB4T1, as subcutaneous tumors. In culture, MB4 cells are large and fibroblastoid, and at high density form an interwoven mat many cells thick. They have high levels of alkaline phosphatase, high ratios of N-acetyl-hexosaminidase to β-glucuronidase (Hex/β-Gluc) (Sherman, 1975a), and do not have detectable levels of Endo antigen (K. Artzt, personal communication), a marker found on parietal yolk sac carcinomas and certain other endodermal and mesodermal cell lines. (K. Artzt, H. Jakob, and F. Jacob, in preparation). MB4T1 tumors (Fig. 1d) are composed of a mixture of two cell types, one highly vacuoated with abundant foamy cytoplasm and the other considerably smaller (M. Sherman, R. Miller, and C. Richter, in preparation). MB2 cultures are mixtures of triangular foamy cells and round, lightly attached cells. The round cells become relatively more abundant at high population densities. MB2 cultures do not produce high levels of alkaline phosphatase or high ratios of Hex/β-Gluc, but do exhibit Endo antigen (K. Artzt, personal communication). MB2T1 tumors (Fig. 1b) are composed of areas of large polygonal cells, sometimes multinucleate, with considerable variations in size and nuclear/cytoplasmic ratio. Clusters of cells are often surrounded by bands of extracellular eosinophilic secretion resembling the Reichert's-membrane-like material produced by parietal yolk sac carcinomas (Pierce

---

ABBREVIATIONS AND TERMINOLOGY

Hex/β-gluc - ratio of specific activities of N-acetyl hexosaminidase to β-glucuronidase.
DVME - Dulbecco-Vogt modified Eagle's medium.
HBSS - Hanks' balanced salt solution.
SSF_1 - F_1(SWR/J x SJL/J).

The day of observation of the sperm plug is considered the first day of pregnancy.

et al., 1962). The initial MB2T1 tumor, shown in Fig. 1b, also contains giant cells resembling trophoblast, but these could no longer be found in subsequent transplant generations.

Although our studies of these and other blastocyst-derived lines suggest similarities between the lines and specific portions of the early mouse conceptus (Sherman, 1975a), additional markers would be useful in clarifying such relationships. In order to provide new reagents for distinguishing among cultured cell lines, to study similarities between these lines and embryonic tissues, and to seek new insights into epigenetic changes in the embryo during differentiation *in vivo* and *in vitro*, we have produced a syngeneic antiserum directed against a blastocyst-derived line, MB4. We report here our initial characterizations of this serum, with emphasis on the presence of cross-reacting antigens in

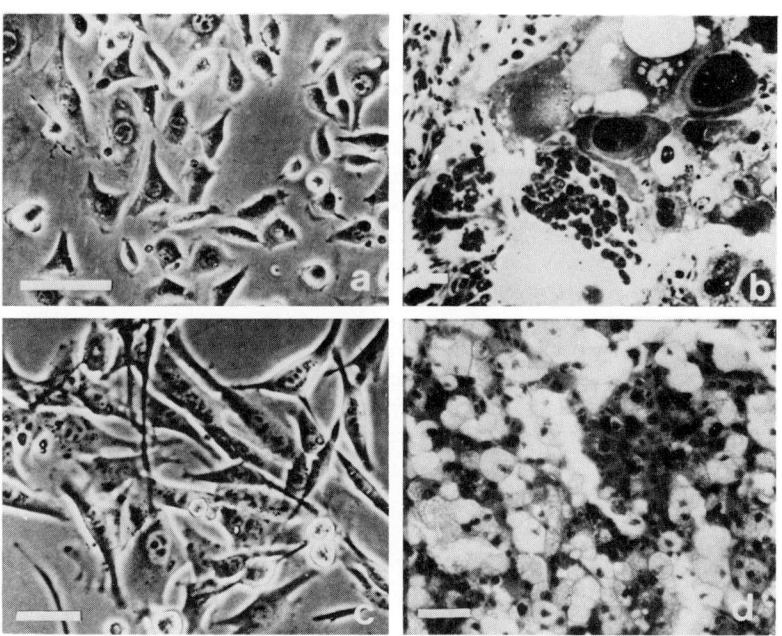

Figure 1. Morphology of blastocyst-derived cell lines and their malignant derivatives: a) MB2 cells in culture, phase contrast; b) MB2T1 primary tumor, hematoxylin and eosin-stained section; c) MB4 cells in culture, phase contrast; d) MB4T1 tumor, hematoxylin and eosin-stained section. The bars represent 50 um.

cultured cell lines derived from teratomas, blastocysts and tumors of differentiated mouse tissues.

EXPERIMENTAL PROCEDURES AND RESULTS

Cell Lines

MB2, MB4, MB21, and MB31 have been previously described (Sherman, 1975a,b); they are derived from cultured (SWR/J x SJL/J)F_1 (SSF_1) blastocysts. MB6 is an independent cell line derived in the same way. MB2T1 and MB4T1 are malignant derivatives of their respective parental lines (Sherman, Miller and Richter, in preparation). MB4C2A and MB4C2M are subclones of MB4. Hepa-1a was derived from the C57L/J hepatoma BW7756 by Dr. G. Darlington (see Bernhard et al., 1973). All of the above cell lines are maintained in MAB 87/3 medium, supplemented with 10% fetal calf serum and antibiotics.

A9, a subline of Earle's L cells, was obtained from the American Type Culture Collection. FBU, a derivative of Friend's erythroleukemic clone 745, was given to us by Dr. J.-F. Conscience. TSD-4 is a nonmalignant epithelial derivative of the 129/Sv embryo-derived teratocarcinoma, OTT 6050. F9 and PCC4.aza1 are, respectively, nullipotent and pluripotent embryonal carcinomas (Artzt et al., 1973 and Jakob et al., 1973) and were given to us by Drs. T. Boon and E. Vitteta, respectively. These lines were maintained on Dulbecco-Vogt modified Eagle's medium (DVME) with 10% fetal calf serum and antibiotics. All cells were grown in Falcon plastic flasks at 37° under 10% CO_2 in air, and were passaged with a solution of Viokase in a Ca^{++} and Mg^{++}-free balanced salt solution. All these lines have been shown to be free of PPLO contamination during the course of this study.

Immunization Procedures

Cells were harvested from monolayer culture with a 0.02% EDTA solution in phosphate-buffered saline, washed twice by centrifugation in Hanks balanced salt solution (HBSS), and suspended in an equal volume of HBSS prior to injection. To produce the anti-MB4 serum, SSF_1 mice, of both sexes, aged 6-10 weeks at the time of the first injection, were given 0.1 ml of cell suspension intraperitoneally at seven-day intervals. The mice were bled from the tail 7 days after the 8th injection. After a 19 day hiatus a booster was administered, and a second set of bleedings taken 6 and 9 days thereafter. Sera from a given bleeding were pooled, heat-inactivated at 56° for 30 min, and stored in aliquots at -70°.

In our immunofluorescence assay, cells harvested without the use of enzymes are washed with DVME, incubated at 4° with anti-MB4 or control serum diluted 1:5 with DVME, washed three times to remove unbound mouse antibody and then incubated with fluorescein-conjugated rabbit anti-mouse IgG immunoglobulin (2 mg/ml protein). After another three washes, the cells are examined by fluorescence microscopy. To provide an objective record of relative fluorescent intensities, experimental and control cells are photographed by equal timed exposures, and the resulting negatives are identically printed.

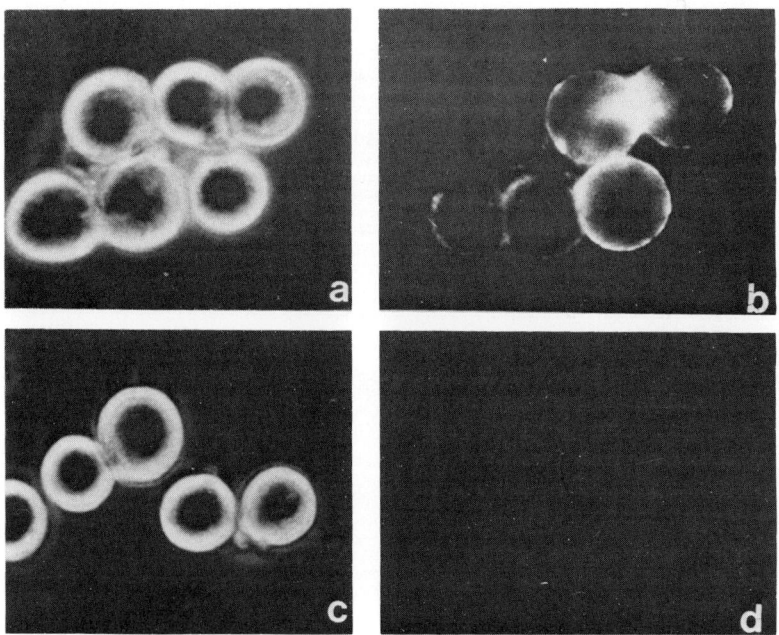

Figure 2. Indirect immunofluorescent reaction of MB4 cells with SSF_1 antiserum to MB4 cells (a and b) and with normal, nonimmune SSF_1 serum (c and d). Panels a and c are phase contrast micrographs; b and d are fluorescence micrographs of these two fields, each taken with an exposure time of 30 sec and then processed together and identically.

Reaction of Anti-MB4 Serum with MB4 Cells and Other Cell Lines

Figures 2b and 2d show that our anti-MB4 serum stains MB4 cells, by indirect immunofluorescence, far more than does nonimmune serum from mice of identical genotype. In most experiments, preincubation with normal SSF_1 serum produces either no fluorescence (e.g. Fig. 2d) or a dim surface fluorescence easily distinguished from the bright ring reaction produced by preincubation with hyperimmune serum. A second control serum was obtained from SSF_1 mice multiply injected with another blastocyst-derived cell line, MB2; this serum, like normal sera, fails to produce surface fluorescence on MB4 cells. MB4 cells do not bind detectable amounts of fluorescein-conjugated rabbit anti-mouse IgG immunoglobulin.

We next examined a number of other cultured lines for their ability to bind anti-MB4 antiserum. Table 1 summarizes the results of these studies. Among the blastocyst-derived lines, positive reactions were obtained with MB21, MB31, and MB6, with two morphologically distinguishable clones (MB4C2A and MB4C2M) of MB4 and with its malignant

TABLE 1

DETECTION OF MB4 ANTIGENS ON CULTURED CELL LINES

Cell Line	Source	MB4 Antigen
MB4		+
MB21	4th-day blastocysts of	+
MB31	SSF_1 mice	+
MB6		+
MB4T1	MB4	+
MB4C2A	MB4	+
MB4C2M	MB4	+
MB2	4th-day blastocysts of SSF_1 mice	−
MB2T1	MB2	−
F9	129/Sv teratocarcinoma	−
PCC4.aza1	OTT 6050	−
TSD-4		+
A9	C3H/HeJ fibrosarcoma	+
FBU	DBA/2J erythroleukemia	+
Hepa-1a	C57L/J hepatoma	+

variant, MB4T1. One blastocyst-derived line, MB2, and its malignant derivative MB2T1, consistently failed to bind anti-MB4 antiserum.

Four kinds of teratoma-derived cells were also tested, all of which originated in the embryoid body ascitic form of the six-day embryo-derived teratocarcinoma OTT 6050 (Stevens, 1970). Two embryonal carcinoma cultures, the nullipotent F9 and the pluripotent PCC4.azal, were negative for cross-reacting determinants. TSD-4, however, a line of nonmalignant epithelial cells derived from OTT 6050 embryoid bodies, was able to bind anti-MB4 antiserum. Embryoid bodies themselves, washed and tested directly after removal from ascitic fluid without prior culture, also failed to exhibit surface immunofluorescence.

We wished to rule out the possibility that these unreactive cells might be coated with secreted material which could prevent access of immunoglobulins to the membrane itself. We showed that this is not the case, at least for F9, by obtaining a positive immunofluorescent reaction with a syngeneic anti-F9 serum (Artzt et al., 1973) given to us by Dr. K. Artzt.

Cross-reacting antigens are also found on several cultured malignant cell lines derived from adult mice, including a hepatoma (Hepa-1a), a fibrosarcoma (A9), and a Friend erythroleukemia (FBU).

Properties of Absorbed Anti-MB4 Serum

Because our initial interest in this syngeneic antiserum stemmed from our need for reagents which could distinguish among cells of different tissues, we wished to see if anti-MB4 serum would retain activity towards MB4 cells after absorption with positive cells likely to differ in epigenetic status. Accordingly, we have tested anti-MB4 serum diluted 1:5 and then absorbed twice with A9 cells (volume ratio of serum to cells was 2:1), for residual staining activity towards these two cultures. Although our fluorescent assay does not easily lend itself to quantitation, we have observed in three separate experiments that absorption by A9 cells leaves a serum which has lost nearly all affinity for A9, but which retains considerable activity towards MB4. Unabsorbed serum stains A9 more brightly than it does MB4; A9-absorbed serum has lost a little of its affinity for MB4, but nearly all its ability to coat A9 cells (Fig. 3).

Is Anti-MB4 Serum Directed Against a Plasma Membrane Molecule?

To determine whether the antigenic components of MB4

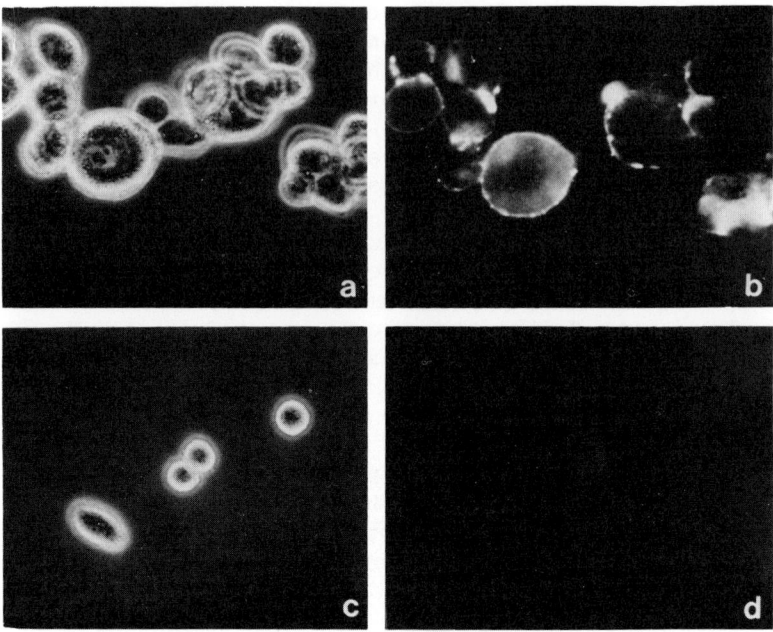

Figure 3. Indirect immunofluorescent reaction of A9-absorbed anti-MB4 serum with MB4 cells (a and b) and with A9 cells (c and d). Panels a and c are phase contrast micrographs; panels b and d are fluoresence micrographs of the same fields, identically exposed and processed. The two A9 cells furthest towards the right showed faint residual fluorescence.

cells are truly elements of the plasma membrane, rather than some secreted material, we have exposed fluorescent-stained MB4 and A9 cells to conditions which induce aggregation and 'capping' of known membrane macromolecules, such as IgG on splenic lymphocytes (Taylor et al., 1971) and H-2 on thymocytes (Unanue et al., 1972). After our immunofluorescence procedure was carried out on MB4 or A9 cells at 4°, cells were incubated at 37° and aliquots examined at intervals. The resultant changes in antigenic distribution are illustrated in Figure 4. Immediately after staining, fluorescence is distributed homogeneously, or in a fine-grained pattern, on the surface of both cell types. With increasing time at 37°, the pattern on many cells alters to one characterized by larger, brighter grains, less numerous than before, which are presumably aggregates of the originally-dispersed antigenic areas. After 30 min of incubation, these large grains will frequently have collected into one or several patches on the cell surface. Cap-like accumulations of antigen-antibody complexes at one pole are also seen, but only in 10-30% of the cells. A9 cells cap much more frequently than do MB4 cells. This pattern of antigen redistribution differs from the fast and nearly complete capping reaction of IgG-bearing splenic lymphocytes (Taylor et al., 1971), but is reminiscent of the slow aggregation of H-2 sites on mouse thymocytes (Unanue et al., 1972).

DISCUSSION AND CONCLUSIONS

Our principal result is that (SWR/J x SJL/J)F_1 mice can produce antibodies in response to determinants found on syngeneic, blastocyst-derived cultured cells. Theoretically, one would expect such a serum to recognize only determinants to which the mouse immune system was not exposed during the development of self-tolerance. Such determinants might be a) present in immunologically protected sites in adult mice, or b) found only in cultured cells of indefinite lifespan, or c) transiently present in early embryonic material. In this last case, the antigen might be characteristic of many different embryonic types, or restricted to a particular tissue or set of tissues.

As Table 1 demonstrates, MB4 shares at least some antigens with cells derived from many different tissues and several strains of mouse. Two cell types, however, embryonal carcinomas and derivatives of MB2, do not express antigens cross-reacting with MB4. The MB4 antigens, therefore, are not universally present on cultured cell lines of indefinite lifespan, and are not artifacts of *in vitro* manipulation, but must to some extent be cell-type specific. In other words, differential expression of the genes responsible for MB4 antigen expression suggests that the gene products may be of biological significance.

That A9-absorbed anti-MB4 serum retains much of its ability to bind to MB4 surface components implies the presence of at least two classes of MB4 antigens, only one of which is also found on A9 cells. We have yet to determine which of these produces the staining reactions of the other positive blastocyst-derived (MB6, MB21, MB31) and non-blastocyst-derived (Hepa-1a, FBU, TSD-4) permanent cell lines. Experiments are in progress to determine whether either of these MB4 antigens can be detected on specific embryonic tissues or on primary cultures derived from these tissues.

Although MB4 antigen was not detectable on either of two embryonal carcinoma lines tested (F9 and PCC4.aza1), it was

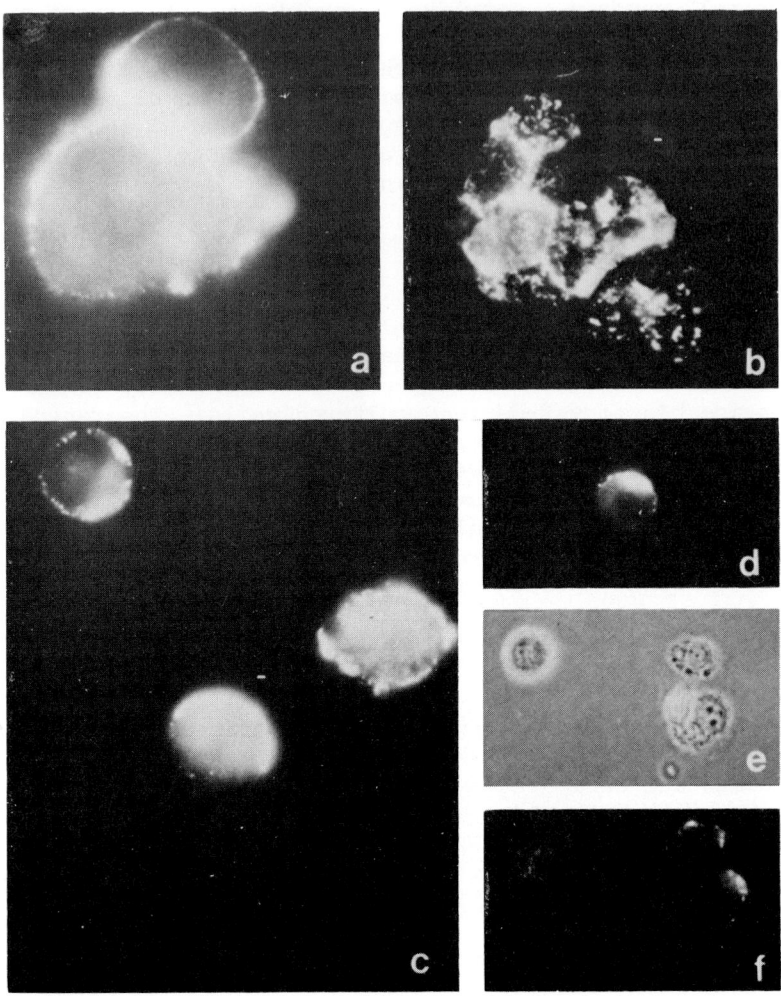

present on TSD-4, a nonmalignant epithelioid derivative of cultured OTT 6050 embryoid bodies. If the progression of teratoma cell types which occurs *in vitro* is a good model of embryonic events, we would not expect to see cross-reacting antigens on morulae or blastocysts, but might hope to detect these determinants on some, or perhaps many, different post-implantation tissue types.

We can say little at present about the function(s) of the molecules our serum defines, or about their relationships to other antigens. The ability of antibody-crosslinked sites to aggregate at 37° suggests that the antigens are integral parts of the plasma membrane, and not a secreted portion of

Figure 4. Redistribution of antigen-antibody complexes after various lengths of time at 37°.
a) MB4 cells immediately after completion of staining procedure. Note numerous fluorescent patches homogeneously distributed over both cells.
b) MB4 cells after 45 min at 37°. Many of these cells show a characteristic coarse-grained pattern.
c) A9 cells after 25 min at 37°. The cell in the center is capped; the one at the left is partially so.
d) Another capped A9 cell after 30 min incubation.
e and f) Phase contrast and fluoresence micrographs of capped A9 cells, partially squashed by the coverslip. Note that in both capped cells, antigen-antibody complexes have collected over the nucleus and opposite most of the cytoplasm. This form of redistribution resembles that characteristic of thymocytes, as distinct from the behavior of splenocytes, in which caps usually form over the Golgi area (Stackpole et al., 1974).

any glycocalyx layer; this question needs further investigation. MB4 antigen is clearly not identical to the F9 antigen identified by Artzt et al. (1973), since F9 and PCC4.azal bear only the latter. A second antigen characteristic of teratoma-derived cells, Endo (Artzt, Jakob and Jacob, in preparation), also differs from the molecules we are studying, in that Endo antibody can be absorbed by MB2, but not by MB4 cells (K. Artzt, personal communication).

Preliminary data (B. Knowles, personal communication) suggests that MB4 cells may contain antigens (p30 and gs3; see Lilly and Steeves, 1974) characteristic of C-type RNA tumor viruses, but that these antigens cannot be detected in MB2 cells. We must therefore consider three possibilities: 1) anti-MB4 serum reacts only with viral-related antigens (though presumably not p30 or gs3, to which mice are tolerant--see Huebner et al., 1971, and Nowinski et al., 1974); 2) anti-MB4 serum does not contain anti-viral antibodies, in which case the relationship found so far between MB4 antigen expression and endogenous virus could be either coincidental or due to some event(s) leading both to MB4 antigen expression and viral protein production; 3) anti-MB4 serum contains antibodies to both viral-associated and viral-independent antigens.

We cannot yet rigorously eliminate any of these three hypotheses, but we believe that our absorption experiments render the first possibility unlikely. L cells make large amounts of viral antigens, yet A9-absorbed serum retains considerable anti-MB4 staining activity. Furthermore, we have been unable to demonstrate any anti-G_{IX} activity (see Stockert et al., 1971) in our unabsorbed MB4 serum by indirect immunofluorescence on 129/J thymus cells (unpublished observations). It is, however, still possible that our antiserum might contain reactivities to other viral-associated determinants. The first of the three hypotheses can be ruled out by finding a cell type which is MB4-reactive but unreactive with anti-viral antibodies; we plan to screen MB4-positive cells for gs3 antigens with this problem in mind. It will also be helpful to see if anti-MB4 serum has any virus-neutralizing activity.

We believe that further study of this and other syngeneic mouse antisera will provide increasingly precise ways of detecting analogies between embryonic tissue types on the one hand, and teratoma or blastocyst-derived cultured cell lines, on the other, and may give us increasingly useful probes for the analysis of developmental transitions *in utero* and *in vitro*.

ACKNOWLEDGEMENTS

Karen Artzt has provided expert advice, Mae Reger and Susanne Anderson secretarial skill, and Vincent Salerno technical assistance; we are grateful to all four. Financial support for this work was provided by USPHS GM 09966; R.A.M. is a predoctoral trainee of the Medical Scientist Training Program.

REFERENCES

Artzt, K., Dubois, P., Bennett, D., Condamine, H., Babinet, C., Jacob, F. (1973). Surface antigens common to mouse cleavage embryos and primitive teratocarcinoma cells in culture. Proc. Nat. Acad. Sci. U.S. 70, 2988.

Artzt, K., Bennett, D., Jacob, F. (1974). Primitive teratocarcinoma cells express a differentiation antigen specified by a gene at the T-locus in the mouse. Proc. Nat. Acad. Sci. U. S. 71, 811.

Bernhard, H. P., Darlington, G. J. Ruddle, F. H. (1973). Expression of liver phenotypes in cultured mouse hepatoma cells: synthesis and secretion of serum albumin. Develop. Biol. 35, 83.

Gluecksohn-Waelsch, S., Erickson, R. P. (1970). The T-locus of the mouse: Implications for mechanisms of development. Curr. Topics Develop. Biol. 5, 281.

Huebner, R. J., Sarma, P. S., Kelloff, G. J., Gilden, R. V., Meier, H., Myers, D. D., Peters, R. L. (1971). Immunological tolerance to RNA tumor virus genome expression: Significance of tolerance and prenatal expressions in embryogenesis and tumorigenesis. Ann. N. Y. Acad. Sci. 181, 246.

Jakob, H., Boon, T., Gaillard, J., Nicolas, J.-F., Jacob, F. (1973). Teratocarcinome de la souris: isolement, culture, et proprietes de cellules a potentialites multiples. Ann. Microbiol. (Inst. Pasteur) 124B, 269.

Kahan, B. W., Ephrussi, B. (1970). Developmental potentialities of clonal *in vitro* cultures of mouse testicular teratoma. J. Nat. Cancer Inst. 44, 1015.

Lilly, F., Steeves, R. (1974). Antigens of murine leukemia viruses. Biochim. Biophys. Acta 355, 105.

Nairn, R. C. (Ed.) (1969). "Fluorescent Protein Tracing". Third edition. Livingstone, London.

Nowinski, R. C., Kaehler, S. L., Burgess, R. R. (1974). Immune response in the mouse to endogenous leukemia viruses. Cold Spring Harbor Symp. Quant. Biol. 39, 1123.

Pierce, G. B., Jr., Midgley, A. R., Jr., Sri Ram, J., Feldman, J. D. (1962). Parietal yolk sac carcinoma: clue to the histogenesis of Reichert's membrane of the mouse embryo. Amer. J. Path. 41, 549.

Rosenthal, M. D., Wishnow, R. M., Sato, G. H. (1970). *In vitro* growth and differentiation of clonal populations of multipotential mouse cells derived from a testicular teratocarcinoma. J. Nat. Cancer Inst. 44, 1001.

Sherman, M. I. (1975a). Long term culture of cells derived from mouse blastocysts. Differentiation 3, 51.

Sherman, M. I. (1975b). The culture of cells derived from mouse blastocysts. Cell 5, 343.

Stackpole, C. W., Jacobson, J. B., Lardis, M. P. (1974). Two distinct types of capping of surface receptors on mouse lymphoid cells. Nature (London) 248, 232.

Stevens, L. C. (1970). The development of transplantable teratocarcinomas from intratesticular grafts of pre- and postimplantation mouse embryos. Develop. Biol. 21, 364.

Stockert, E., Old, L. J., Boyse, E. A. (1971). The G_{IX} system. A cell surface allo-antigen associated with murine leukemia virus; implications regarding chromosomal integration of the viral genome. J. Exp. Med. 133. 1334.

Taylor, R. B., Duffus, W. P. H., Raff, M. C., de Petris, S. (1971). Redistribution and pinocytosis of lymphocyte

surface immunoglobulin molecules induced by anti-
immunoglobulin antibody. Nature New Biol. 233, 225.
Unanue, E. R., Perkins, W. D., Karnovsky, M. J. (1972).
Ligand-induced movement of lymphocyte membrane macromolecules. I. Analysis by immunofluorescence and ultrastructural radioautography. J. Exp. Med. 136, 885.

IV.
Teratoma-Host Interactions

CONTROL OF TERATOCARCINOGENESIS

DAVOR SOLTER[1], NANCY ADAMS[1], IVAN DAMJANOV[2]
AND HILARY KOPROWSKI[1]

[1]The Wistar Institute of Anatomy and Biology
36th Street at Spruce
Philadelphia, Pennsylvania 19104
and [2]Department of Pathology
University of Connecticut Health Center
Farmington, Connecticut 06032

This paper deals with the control of teratocarcinogenesis in mice. The tumors we will be dealing with can be divided into either benign (mature) teratoma or malignant teratoma (teratocarcinoma). We use the term *teratocarcinogenesis* to mean the appearance (production, induction) of both benign and malignant teratomas. Both can arise spontaneously or can be experimentally induced, but we may assume that control of teratocarcinogenesis is not entirely the same in both cases.

It is now commonly assumed that teratomas and teratocarcinomas develop from a multipotential or totipotent stem cell, either a primordial germ cell or a cell from an early embryo (Pierce, 1967; Stevens, 1967; Damjanov and Solter, 1974a). In the normal course of events, such a cell would develop and differentiate into a spermatozoon or oocyte or into any of the somatic cells in the adult organism. It is of great importance to establish what diverts the undifferentiated cell that then becomes the stem cell of a malignant tumor. Studies of this "disease of cell differentiation" (Markert, 1968) might not only explain teratocarcinogenesis but also reveal one of the basic mechanisms of neoplasia.

SPONTANEOUS TERATOCARCINOGENESIS

In commonly used laboratory animals, spontaneous gonadal and extragonadal teratomas are rare (for reviews, see Stevens, 1967; Damjanov and Solter, 1974a). However, the discovery of a mouse strain with a relatively high incidence of spontaneous testicular teratomas (Stevens and Little, 1954) has made it possible to study the controls involved in teratocarcinogenesis.

Genetic Control

Spontaneous testicular teratomas were first described by Stevens and Little (1954) in the testes of the inbred 129 mouse strain; about 1% of all animals examined had teratomas. So far, extensive search has failed to reveal any other mouse strains with a similar incidence. Slye et al. (1919), for example, did not find a single testicular teratoma in 19,000 autopsies of male mice and only three cases of spontaneous testicular teratomas have been described in strains other than 129 (Stevens, 1967; Meier et al., 1970). Stevens and Mackensen (1961) investigated different sublines of strain 129 and noted that the incidence of teratomas ranged from 0.26% to 1.70% suggesting the presence of multiple factors (genetic?) regulating teratocarcinogenesis. This suggestion was corroborated when they found almost no teratomas in F_1 hybrids and backcrosses between 129 and several other strains (1 teratoma out of 11,292 males examined). Stevens and Mackensen (1961) also investigated the effect of introducing different mutant genes into the 129 background. While most of the mutant genes tested did not alter the incidence of teratomas, the gene Steel (Sl^J) considerably increased the incidence (to almost 7%) in heterozygous males. The Sl^J gene affects the development of pigment, red blood and germ cells (Bennett, 1956). In heterozygous animals during prenatal development, the numbers of germ cells are reduced but it is not clear whether the teratocarcinogenic action of the Steel gene is related to its action on germ cell development.

Recently, Stevens (1973) described a new inbred subline of the 129 strain, the 129/ter Sv, in which the incidence of testicular teratomas in males was approximately 30%. This very high incidence is probably due to a single gene mutation

---

ABBREVIATIONS AND TERMINOLOGY

ATS - rabbit anti mouse thymocyte serum.
ALS - rabbit anti mouse lymphocyte serum.

The day of observation of the sperm plug is considered day zero of pregnancy.

since hybrids between 129/ter Sv and other 129 sublines have an intermediate incidence. Another example of genetically controlled teratocarcinogenesis is the LT mouse strain which has a high incidence (about 50%) of ovarian teratomas (Stevens and Varnum, 1974). At present, it is not clear whether only a single gene is involved in this case. Ovarian teratomas were found in less than 10% of F_1 hybrids between LT and C57BL/6 mice; either very few genes are involved in the control of teratocarcinogenesis or the penetrance of a single gene is altered in an F_1 background.

From these results we can conclude that genetic factors play an important, if not decisive, role in spontaneous teratocarcinogenesis in mice. The exact mechanisms are not known although apparently, in both ovarian and testicular teratocarcinogenesis, the initial change is the induction of something akin to normal embryogenesis. This is more obvious in ovarian teratomas which are actually derived from parthogenetically activated ova which develop almost normally up to the egg-cylinder stage but then become disorganized and give rise to teratomas. The percentage of teratocarcinomas in ovarian and testicular tumors is relatively small--approximately 20% in ovarian tumors based on the presence of undifferentiated cells (Stevens and Varnum, 1974). From the available data it is not clear whether there is any genetic influence in determining if a spontaneous tumor will develop into a mature teratoma or teratocarcinoma.

Environmental Control

Several factors affect teratocarcinogenesis in the genetically susceptible 129 strain of mice. The incidence of tumors is two to three times higher in the left testes than in the right (Stevens and Mackensen, 1961; Stevens, 1973). Ovarian teratomas were observed more often in the right ovary than in the left (Stevens and Varnum, 1974). Whether these differences are consequent to some trivial anatomical or developmental variation or whether they are related to the problem of teratocarcinogenesis is, at present, unclear. It seems, however, that some animals are more susceptible to ovarian or testicular teratoma since the incidence of bilateral tumors is much higher than one would expect if the incidence of teratocarcinogenesis in the left and right gonad were completely independent.

The incidence of teratomas in males of second and later litters was about twice that of the first litter (Stevens and Mackensen, 1961). Changes in the maternal environment after the first parturition, probably in their immune status with respect to fetal antigen(s), might explain these observations.

Attempts to increase the incidence of teratomas have been singularly unsuccessful. Hormones such as testosterone, estradiol, progesterone, cortisone, etc., carcinogens and X-irradiation have been applied without effect to pregnant females and newborn male mice of the 129 strain (Stevens and Mackensen, 1961). Bresler (1959) reported that injection of metallic salt could induce teratomas in mouse testes but Stevens (1967) was unable to reproduce these results in strain 129 mice. Hormones, however, could be involved in the appearance of ovarian teratomas since the first tumors were observed in females 2-months-old, i.e., sexually mature. Since tumors are derived from parthenogenetically activated ova, oocyte maturation might be necessary for the initiation of parthenogenesis.

EXPERIMENTAL TERATOCARCINOGENESIS

Experimental teratomas can be produced in mice in two ways. One is the grafting of genital ridges isolated from mouse embryos into adult testes (Stevens, 1964); the other is the transplantation of mouse embryos into the testis (Stevens, 1968, 1970c) or under the kidney capsule (Solter et al., 1970) of an isogenic host. The ease and effectiveness of both these procedures greatly facilitate the elucidation of those factors influencing the incidence and the nature of developing tumors.

Teratomas Derived from Genital Ridges

Genetic control

Genital ridges isolated from fetuses and transferred to the testes of adults develop into ovaries or testes depending on the sex of the donor fetus. Teratomas were observed only within the testis grafts (Stevens, 1964) and they were histologically similar to spontaneous teratomas. However, the incidence of teratomas in testis grafts was considerably higher (approaching 100%) when suitable combinations of different regulatory factors were used (see below). Teratomas also readily developed in grafted genital ridges of the A/He mouse strain and of 129 X A/He F_1 hybrids although spontaneous teratomas were extremely rarely observed in A/He strains (Stevens, 1967; Meier et al., 1970) and never in 129 X A/He F_1 hybrids (Stevens, 1970a). Apparently, the grafting procedure considerably enhances teratocarcinogenesis in the genetically susceptible 129 strain and also induces teratomas in strains where spontaneous tumors are never observed. However, the whole process of teratocarcinogenesis still appears to be under genetic control as there are mouse strains in which very few grafted genital ridges develop into teratomas (Stevens, 1970a) and some, like C3H, whose genital ridges are completely resistant to teratocarcinogenesis when transplanted (Stevens, 1970c). It would be of considerable interest to investigate the development of grafts obtained from F_1 hybrid fetuses between highly susceptible and highly resistant mouse strains.

Age of the graft

Stevens (1964, 1966, 1970a) showed the importance of the age of the grafted fetal genital ridges in experimentally induced teratocarcinogenesis. Teratomas developed in about 80% of the grafted genital ridges if they were isolated from 11 1/2- and 12 1/2-day-old embryos; the incidence dropped sharply (to less than 20%) when the embryos were 13 to 13 1/2 days old and to less than 10% in 14-day-old and older embryos. Since the incidence of spontaneous teratomas is 8% we can conclude that the process of grafting genital ridges to testes is not teratocarcinogenic if the genital ridges are isolated from 14-day-old or older embryos.

It also seems that the age-dependent susceptibility to teratocarcinogenesis is genetically controlled. The highest incidence of teratomas in the 129 mouse strain occurs when genital ridges removed from 11- and 12-day-old embryos are grafted. The incidence in A/He strain for embryos of the same age is considerably lower, increasing when the genital ridges from 13-day-old embryos are used. The incidence of teratomas in transplanted genital ridges of 129 X A/He F_1 hybrid embryos is higher than in any of the parental strains (Stevens, 1970a) and an additive effect is also obvious from the fact that a high incidence of teratomas is present in grafts removed on day 12 (as in the 129 strain) and on day 13 (as in the A/He strain).

Site of the graft

Susceptible genital ridges of 129, A/He and their F_1 hybrids were grafted to different sites in male and female hosts (Stevens, 1970b). The incidence of teratomas was five to ten times higher in grafts made to scrotal sites (testis, epididymis, epididymal fat pad) than in those grafted to non-scrotal sites (spleen, liver, kidney, ovarian fat pad, etc.).

The incidence was also higher in genital ridges grafted to normal testes than to cryptorchid testes. These results indicate that the scrotal site exerts a strong teratocarcinogenic influence. This influence may result from several sources: one is probably the lower temperature in the scrotum; another may be directly related to the testis since of all scrotal sites the highest incidence occurs in the testis itself; and finally the incidence in cryptorchid testes is higher than in other nonscrotal sites.

The results also suggest that grafting per se has some teratocarcinogenic influence regardless of the site (Stevens, 1970b). For example, in 129 strain mice the incidence of teratomas in genital ridges grafted to nonscrotal sites is the same as that occurring spontaneously and since the spontaneous incidence is quite high, the effect of grafting is probably masked. However, in the A/He strain where spontaneous teratomas are never found, grafting to nonscrotal sites still resulted in a 3-4% incidence of teratomas (Stevens, 1970b). What component of the grafting procedure is responsible for the teratocarcinogenic effect is unknown; it may be the handling itself, the short exposure to room temperature, the absence of controlling influences on the genital ridge after removal from the fetus, or some other effect.

Other factors

The laterality of the genital ridge and the seriation of the litter from which the genital ridges were removed did not affect the incidence of teratomas (Stevens, 1964) although laterality and seriation exerted an influence on spontaneous teratocarcinogenesis (see above). It is quite probable that effects which are obvious in low incidence of spontaneous teratoma might become masked in the high incidence of teratomas in grafted genital ridges. Castration and the administration of sex hormones did not change the incidence of teratomas, suggesting again that sex hormones are probably not involved in experimental teratocarcinogenesis.

Embryo-Derived Teratomas

The problems of the control of teratocarcinogenesis in grafted embryos are somewhat different from those we have been discussing. Grafted embryos, if they grow at all, always develop into teratomas; therefore, the incidence of teratomas in recovered grafts is always 100%. A much more interesting question, therefore, is what percentage of tumors contain undifferentiated stem cells and can be classified as teratocarcinomas. Another somewhat related question is what percentage of embryos can develop in extrauterine sites in relation to the age of the grafted embryo.

Genetic control

Stevens (1968; 1970c) grafted 1-, 2- and 4-8-cell embryos and blastocysts into testes of adult mice, and examined them for up to 60 days after grafting for the presence of extraembryonic and embryonic derivatives and for the presence of undifferentiated embryonic cells. The percentage of recovered grafts was relatively small, ranging from 1% for zygotes to as high as 20% for blastocysts. The grafts containing undifferentiated embryonic cells on the basis of morphological criteria will be considered as teratocarcinomas in this discussion since it has been repeatedly shown that only those tumors with undifferentiated stem cells can proliferate and be indefinitely retransplanted (Pierce, 1967; Stevens, 1967).

Data on genetic influences is available only for grafts of 2-cell embryos (Stevens, 1968). Embryos of the 129 strain and 129 x A/He F_1 hybrids developed into teratomas of which some contained undifferentiated embryonic cells. Embryos from other strains (A/He and several substrains of C57BL) did not develop at all (Stevens, 1968). Grafts were recovered from 20% of the transplanted 129 blastocysts; about one-third of these could be classified as teratocarcinomas. Blastocysts from 129 x A/He F_1 hybrids developed in 10% of all transfers and about half of the grafts had undifferentiated embryonic cells. All these data indicate that genetic control might determine both the embryo survival and development and also the proportion of teratoma and teratocarcinoma in resulting tumors, but a much larger study of teratomas derived from preimplantation embryos is certainly indicated.

More than 70% of 6-day-old embryos of strain 129 and 129 X A/He F_1 hybrids developed into tumors of which more than half had undifferentiated embryonic cells when examined one month after grafting (Stevens, 1970c). Tumors developed from about 40% of 6-day-old A/He embryos and half of the tumors contained undifferentiated embryonic cells (Stevens, 1970c). It is difficult to say whether the percentage of recovered tumors indicates a real genetic difference between 129 and A/He strain. Our own experience suggests that unsuccessful grafting most probably results from some sort of technical failure and the experiments mentioned above should be repeated in such a way that the positioning of healthy, undamaged embryos can be observed before any conclusions about strain differences in graft recovery are made. When such precautions were observed, we have found that practically all embryos developed into recoverable tumors.

When grafting 7-day-old embryos under the kidney capsule, we have observed one important strain-dependent difference. When embryos of C3H, CBA and A strain were used, about half the tumors contained embryonal carcinoma cells and were classified as teratocarcinomas (Solter et al., 1970; Damjanov and Solter, 1974b,c; Solter et al., 1975). If, however, embryos of the same age but of C57BL and AKR strains were grafted, very few, if any, possessed undifferentiated embryonic cells (Skreb et al., 1972; Damjanov and Solter, 1974b,c; Solter et al., 1975). Table 1 gives our most recent results on the incidence of teratomas and teratocarcinomas in C3H, C57BL and AKR strains. It is necessary to point out that the classification of tumors was based on histological examination and that all tumors containing undifferentiated embryonic cells in the form of embryoid bodies and ectodermal vesicles (see Solter et al., 1975) were classified as teratocarcinomas. Typical nests of embryonal carcinoma cells found in large numbers in C3H tumors were practically never observed in C57BL and AKR tumors. Our results suggest that the nature of embryo-derived tumors, both malignant and benign, is genetically controlled and that teratocarcinoma-permissive and -nonpermissive mouse strains do exist. In contradistinction to the genetic control operating in spontaneous and genital ridge-derived teratoma that was graft related (Stevens, 1964; Stevens and Varnum, 1974), the control of permissiveness or nonpermissiveness for teratocarcinoma in our system is probably host-related (Skreb et al., 1972; Damjanov and Solter, 1974b). Namely, when embryos from a nonpermissive strain are transplanted under the kidney capsule of a F_1 hybrid between a permissive and nonpermissive strain, about half develop into teratocarcinomas (Skreb et al. 1972; Damjanov and Solter, 1974b). Moreover, such teratocarcinomas are retransplantable in F_1 hybrids but almost never in the host of the nonpermissive strain from which the original embryo was isolated (Damjanov and Solter, 1974b). Consequently, we are faced with the basic question stated at the beginning of this article, namely why do some grafted embryos differentiate completely and form benign teratomas, while some indefinitely retain undifferentiated cells and become malignant teratocarcinomas.

When examined shortly after transplantation (7-14 days) *all* grafts, whether permissive, or nonpermissive, contain undifferentiated embryonic cells (Stevens, 1970c; our unpublished data). Therefore, with further time, in some of the

TABLE 1

INCIDENCE OF TERATOMA AND TERATOCARCINOMA AND WEIGHT
OF TUMOR AND SPLEEN IN SEVERAL MOUSE STRAINS

Host	No. of Animals	Teratoma	Teratocarcinoma	Weight[a] Tumor	Spleen
C3H Males	48	16	32	3333 ± 407[d]	363 ± 34[b]
C3H Males-Control	19	--	--	--	114 ± 4
C3H Females	32	13	19	4115 ± 702	357 ± 43[b]
C3H Females-Control	23	--	--	--	162 ± 8
C57BL Males	51	47	4	2032 ± 340 ⎤	342 ± 36[b]
C57BL Males-Control	31	--	--	$p < 0.05$	133 ± 14
C57BL Females	48	45	3	1262 ± 261[c] ⎦	298 ± 34[b]
C57BL Females-Control	34	--	--	--	177 ± 21
AKR Males	44	42	2	1658 ± 234 ⎤	306 ± 29[b]
AKR Males-Control	33	--	--	$p < 0.01$	161 ± 17
AKR Females	42	34	8	3611 ± 482 ⎦	452 ± 35[b]
AKR Females-Control	33	--	--	--	116 ± 8

[a] Weights are given in mg ± standard error of the mean.

[b] Significantly different from controls ($p < 0.01$).

[c] Significantly different from C3H and AKR females ($p < 0.01$).

[d] Significantly different from AKR and C57BL males ($p < 0.01$).

grafts all such cells must differentiate, while in others they persist. These data suggest that all embryos of a certain stage can potentially develop into teratocarcinomas, but that the final nature of a graft depends on its interaction with the host.

Undifferentiated embryonic cells in teratocarcinoma possess specific antigen(s) able to induce an immune response in the syngeneic host (Artzt et al., 1973, 1974). Embryo-derived tumors also induce splenomegaly in the syngeneic host (Table 1) that is much more pronounced if the tumors are teratocarcinomas than if they are teratomas (Damjanov and Solter, 1974c). Splenomegaly is considered to be an immunological reaction in tumor-bearing animals (McKhann and Jagarlamoody, 1971). On the basis of these results we assumed that at least some components of the mechanism which regulates development of grafted embryos into teratoma or teratocarcinoma are immunological in nature. Using C3H and C57BL mice as typical of permissive and nonpermissive strains, we wished to determine whether interfering with the immune system of the host would change the incidence of teratomas and teratocarcinomas in those strains. Preliminary results of such attempts have recently been published (Solter et al., 1975). Splenectomy is supposed to remove the source of blocking antibodies and therefore inhibit tumor growth (Ferrer, 1968; Prehn et al., 1971). Our preliminary results (Damjanov and Solter, 1974c) had seemed to corroborate such a hypothesis, but in the present experiment with C3H mice splenectomized at the time of embryo transfer, the change in the incidence of teratoma vs. teratocarcinoma (Table 2) and the reduction of weight of the tumors was not observed (Table 3). That the use of a different substrain of C3H mice is responsible for this variance seems unlikely but, at present, we have no better explanation. Splenectomy, on the other hand, considerably reduced the weight of tumors in C57BL mice.(Table 5). The incidence of teratoma and teratocarcinoma in both C3H and C57BL mice remained unchanged (Tables 2 and 4). The next step was to try to reduce cellular immunity using anti thymocyte serum (ATS). Our previous results showed, rather unexpectedly, that rigorous ATS treatment (0.25 ml i.p. on days -3, -1, 0, +1, +3 and then twice a week until the animal was sacrificed, which was in all experiments two months after transplantation of the embryo) actually *reduced* the weight of tumors in C57BL mice (Solter et al., 1975). We therefore tried another schedule (0.25 ml i.p. on days 15, 16, 17 after transplantation and then once a week) in the hope that the possible immunostimulative effect of the cellular immune response (Prehn, 1972; Fidler, 1974) would be

TABLE 2

INCIDENCE OF TERATOMA AND TERATOCARCINOMA IN C3H MICE

Host	No. of Animals with Tumor	No. of Animals with Teratoma	No. of Animals with Teratocarcinoma
Males	48	16	32
Females	32	13	19
Females-Ligation of Ovaries	47	21	26
Males-Splenectomized	40	17	23
Females-Splenectomized	31	15	16
Males-Treated with ATS Schedule I[a]	27	15	12
Males-Treated with ATS Schedule II[b]	10	7	3
Females-Treated with ATS Schedule I[a]	27	9	18
Females-Treated with ATS Schedule II[b]	10	5	5

[a]Animals injected with 0.25 ml of rabbit anti mouse thymocyte serum i.p. on days -3, -1, 0, 1, 3 and then twice a week for two months.

[b]Animals injected with 0.25 ml of rabbit anti mouse thymocyte serum i.p. on days 15, 16, 17 after operation and then once a week.

Chi-square analysis showed no significant difference at the 0.05 level.

TABLE 3

EFFECT OF SPLENECTOMY, ATS[a,b] TREATMENT AND LIGATION OF OVARIES ON THE WEIGHT OF TUMORS AND SPLEEN IN C3H MICE

Host	No. of Animals with Tumor (Teratoma/Teratocarcinoma)	Weight[c] Tumor	Spleen
Males	48 (16/32)	3333 ± 407	363 ± 34[e]
Males-Control	19	--	114 ± 4
Males-Splenectomized	40 (17-23)	3309 ± 479	--
Males[a]	27 (15/12)	3038 ± 528	528 ± 58[e]
Males[b]	10 (7/3)	987 ± 462[d]	236 ± 21[e]
Females	32 (13/19)	4115 ± 702	357 ± 43[e]
Females-Control	23	--	162 ± 8
Females-Ligation of Ovaries	47 (21/26)	3180 ± 502	418 ± 37[e]
Females-Ligation of Ovaries-Control	36	--	120 ± 5
Females-Splenectomized	31 (15/16)	3405 ± 654	--
Females[a]	27 (9/18)	2951 ± 603	355 ± 32[e]
Females[b]	10 (5/5)	924 ± 323[d]	253 ± 13[e]

[a] Animals injected with 0.25 ml of rabbit anti mouse thymocyte serum i.p. on days -3, -1, 0, 1, 3 and then twice a week for two months.

[b] Animals injected with 0.25 ml of rabbit anti mouse thymocyte serum i.p. on days 15, 16, 17 after operation and then once a week.

[c] Weights are given in mg ± standard error of the mean.

[d] Significantly different from all other male or female groups ($p < 0.02$).

[e] Significantly different from controls ($p < 0.01$).

TABLE 4

INCIDENCE OF TERATOMA AND TERATOCARCINOMA IN C57BL MICE

Host	No. of Animals with Tumor	No. of Animals with Teratoma	No. of Animals with Teratocarcinoma[d]
Males	51	47	4
Males-Splenectomized	41	41	0
Males-Treated with ATS Schedule I[a]	39	36	3
Males-Treated with ATS Schedule II[b]	10	10	0
Males-Treated with ALS[c]	13	10	3
Females	48	45	3
Females-Ligation of Ovaries	54	49	5
Females-Implantation of Ovaries i.p.	33	23	10
Females-Ovariectomized	32	31	1

$p < 0.05$ between Females and Females-Ligation of Ovaries; $p < 0.01$ between Females-Ligation of Ovaries and Females-Implantation of Ovaries i.p.; $p < 0.01$ between Females-Implantation of Ovaries i.p. and Females-Ovariectomized.

[a] Animals injected with 0.25 ml of rabbit anti mouse thymocyte serum i.p. on days -3, -1, 0, 1, 3 and then twice a week for two months.

[b] Animals injected with 0.25 ml of rabbit anti mouse thymocyte serum i.p. on days 15, 16, 17 after operation and then once a week.

[c] Animals injected with 0.25 ml of rabbit anti mouse lymphocyte serum i.p. on days -3, -1, 0, 1, 3 and then twice a week for two months.

[d] Tumors were classified as teratocarcinoma on the basis of histology. Not all teratocarcinomas have distinct nests of embryonal carcinoma cells but all possess embryoid bodies and ectodermal and endodermal vesicles. Chi-square analysis showed significant difference only where indicated.

TABLE 5

EFFECT OF SPLENECTOMY, ATS[a,b] AND ALS[c] TREATMENT ON THE WEIGHT OF TUMOR AND SPLEEN IN C57BL MICE

Host	No. of Animals with Tumor (Teratoma/ Teratocarcinoma)	Tumor	Weight[d] Spleen
Males	51 (47/4)	2032 ± 340	342 ± 36[f]
Males-Control	31	---	133 ± 14
Males-Splenectomized	41 (41/0)	247 ± 63[e]	---
Males-Treated with ATS Schedule I[a]	39 (36/3)	762 ± 190[e]	267 ± 25[f]
Males-Treated with ATS Schedule II[b]	10 (10/0)	634 ± 202	221 ± 49[f]
Males-Treated with ALS[c]	13 (10/3)	1450 ± 625	496 ± 95[f]

[a] Animals injected with 0.25 ml of rabbit anti mouse thymocyte serum i.p. on days -3, -1, 0, 1, 3 and then twice a week for two months.

[b] Animals injected with 0.25 ml of rabbit anti mouse thymocyte serum i.p. on days 15, 16, 17 after operation and then once a week.

[c] Animals injected with 0.25 ml of rabbit anti mouse lymphocyte serum i.p. on days -3, -1, 0, 1, 3 and then twice a week for two months.

[d] Weights are given in mg ± standard error of the mean.

[e] Significantly different from weight of tumors in untreated males (P<0.01).

[f] Significantly different from control (P<0.01).

preserved at the beginning of tumor development and the detrimental effect of tumor rejection would be decreased. The decrease in the weight of the tumors in C57BL mice was still present and there was no increase in the incidence of teratocarcinoma (Tables 4 and 5). Also, while rigorous treatment with ATS did not change the weight or incidence of teratocarcinomas in C3H mice, the milder ATS treatment reduced the weight and considerably decreased the incidence of teratocarcinomas in the C3H strain (Tables 2 and 3). This was actually the only manipulation we found that could alter the nature of embryo-derived tumors in C3H mice. Splenomegaly observed in controls was also present in animals treated with ATS and it was more pronounced in animals which received rigorous ATS treatment (Tables 3 and 5). Treatment with anti lymphocyte serum (ALS) also reduced the weight of tumors in C57BL males but not as much as ATS administration (Table 5).

All these results suggest an involvement of the host's immune system in the development of embryo-derived tumors. Cellular immunity is apparently a rather important factor in promoting tumor growth. This is corroborated by the fact that both C3H and C57BL embryos grafted under the kidney capsule of congenitally athymic (*nu/nu*) mice developed into very small, benign teratomas (unpublished results).

Environmental factors

Grafting preimplantation embryos, Stevens (1968, 1970c) observed considerable difference in successful recovery of the grafts. While almost none of the grafted zygotes developed into tumors, about 20% of grafted blastocysts did. Granting that technical difficulties are similar in both cases, it would seem that older embryos are more capable of development in extra-uterine sites.

A much higher percentage of postimplantation embryos develop into teratomas, probably partially due to the fact that they are much easier to handle. For us, it was much more important to see if the percentage of teratocarcinomas in grafts is dependent on the age of the embryo. We have shown (Damjanov et al., 1971) that while teratocarcinomas develop in about half of the grafts of 7-day-old embryos, grafting of 8- and 9-day-old embryos always resulted in the development of mature teratomas. Similar results were suggested by Stevens (1970c). Since undifferentiated embryonic cells are present in grafts of 8-day-old embryos if they are examined 7 days after grafting (unpublished results), we suggested that although undifferentiated cells were still present in embryos older than 7 days, they were already determined for specific differentiation and the grafting process was not sufficient for retention of the undifferentiated state. Our results (Damjanov and Solter, 1974c) also showed that C3H-derived tumors grow larger under the kidney capsule than in the testes. However, since the incidence of teratocarcinomas is the same, it is probable that due purely to mechanical reasons, testicular tumors are prevented from reaching very large sizes.

Hormonal influences

From the previous discussion, there appeared to be no hormonal influence on growth of spontaneous testicular and ovarian teratomas or teratomas in grafted genital ridges. We previously reported (Solter et al., 1975) that in C57BL females whose ovaries have been ligated and left to wither in the abdominal cavity, the weight of embryo-derived teratoma was considerably decreased. Hormonal and/or immunological (Coggin and Anderson, 1972) factors could explain such findings, so we performed additional experiments in the hopes of distinguishing between the two. We first realized that the low incidence reported previously (Solter et al., 1975) was probably due to technical difficulties in implanting embryos one week after ligation of the ovaries. When care was taken to make sure that every transplantation was carried out properly, the successful recovery of the grafts was practically 100%. The weight of the tumors, however, was still lower than in control females (Table 6) although the incidence of teratoma vs. teratocarcinoma was not changed. Ligation of the ovaries had similar effects in C3H mice (Tables 2 and 3). Two groups of C57BL females, one ovariectomized and the other nonovariectomized but which had two ovaries deposited in the peritoneal cavity, had embryos transplanted under the kidney capsule one week after these manipulations. The weight of the tumors was significantly decreased in ovariectomized females suggesting that the presence of ovarian hormones, if not essential for tumor growth, might at least stimulate it. This result suggests that examination of the influence of other hormones in embryo-derived teratoma might be fruitful. Another more important finding was the higher incidence of teratocarcinoma in animals with ovaries deposited in the peritoneal cavity (Table 4). As mentioned previously, none of our other procedures has been successful in varying the incidence of teratocarcinomas in the nonpermissive C57BL strain. A possible explanation for the observed increase in teratocarcinoma incidence is that ovarian tissue, disintegrating in the peritoneal cavity, immunized the host animals.

TABLE 6

EFFECT OF LIGATION OF OVARIES, OVARIECTOMY AND IMPLANTATION
OF OVARIES IN THE PERITONEAL CAVITY ON THE WEIGHT OF
TUMOR AND SPLEEN IN C57BL MICE

	No. of Animals with Tumor (Teratoma/ Teratocarcinoma)	Weight[a] Tumor	Spleen
Females	48 (45/3)	1262 ± 261	298 ± 34[b]
Females-Control	34	---	177 ± 21
Females-Ligation of Ovaries	54 (49/5)	987 ± 231	268 ± 34[b]
Females-Ligation of Ovaries-Control	9	--- $p<0.05$	93 ± 18
Females-Implantation of Ovaries i.p.	33 (23/10)	1213 ± 323	292 ± 44[b]
Females-Implantation of Ovaries i.p.-Control	39	---	138 ± 16
Females-Ovariectomy	32 (31/1)	543 ± 194	241 ± 32[b]
Females-Ovariectomy-Control	43	---	113 ± 6

[a] Weights are given in mg ± standard error of the mean.

[b] Significantly different from controls ($p<0.01$).

This changed immunological status of the host either stimulated or allowed undifferentiated cells to remain in this state. Gooding and Edidin (1974) showed that ovaries are the only adult tissues which share some antigenic characteristics with teratocarcinoma cells. If the effect of presumed immunization with ovarian tissues can be verified this would open a whole new approach to the control of teratocarcinogenesis. It might be possible to increase the incidence of teratocarcinoma using immunization with teratocarcinoma or other embryo-derived cells or using pregnant animals as recipients. Work along these lines is now in progress in our laboratory.

CONCLUSION

It is obvious from this discussion that the control of teratocarcinogenesis is still a poorly understood process. First of all, the amount of data for each experimental system is relatively small and the data in most cases merely suggest in what direction one might look for possible control mechanisms. Genetic control is probably present in both spontaneous and experimentally induced teratomas. Moreover, it seems that a genetic influence is necessary, but not a sufficient cause, for appearance of teratomas. On the genetically permissive background some other factors must operate to induce the appearance of teratomas and/or teratocarcinomas in grafted genital ridges and embryos. It is obvious that the site of the graft plays an important role at least in teratocarcinogenesis in transplanted genital ridges. In the case of embryo-derived tumors, removal from the uterus and subsequent disturbance of organized structure is sufficient to produce a teratoma. If, however, the embryo is to develop into a teratocarcinoma, some other control factors are necessary. So far we have identified the importance of the genetic influence of the host but still have little knowledge about how this mechanism of genetic influence operates. The reaction of the host immune system to the presence of undifferentiated embryonic cells and the effect of such reactions on subsequent development of these cells is something which is in our opinion worthy of further investigation.

ACKNOWLEDGEMENTS

The authors' original work was supported by grants CA 04534, CA 10815 and CA 17546 from the National Cancer Institute. One of us (D.S.) was in part supported by a fellowship from the Damon Runyon Memorial Cancer Fund (DRF-810). We thank Elsa Aglow for excellent technical help.

REFERENCES

Artzt, K., Bennett, D., Jacob, F. (1974). Primitive teratocarcinoma cells express a differentiation antigen specified by a gene at the T-locus in the mouse. Proc. Nat. Acad. Sci. U.S. 71, 811.

Artzt, K., Dubois, P., Bennett, D., Condamine, H., Babinet, C., Jacob, F. (1973). Surface antigens common to mouse cleavage embryos and primitive teratocarcinoma cells in culture. Proc. Nat. Acad. Sci. U.S. 70, 2988.

Bennett, D. (1956). Developmental analysis of a mutation with pleiotropic effects in the mouse. J. Morph. 98, 199.

Bresler, V. M. (1959). Experimental teratoids of white mouse testis induced by testosterone and copper sulfate. (in Russian). Vop. Oncol. 5, 24.

Coggin, J. H., Jr., Anderson, N. G. (1972). Phase-specific autoantigens (fetal) in model tumor systems. In "Proceedings of the Second Conference on Embryonic and Fetal Antigens in Cancer" (Eds. N. G. Anderson, J. H. Coggin, Jr., E. Cole and J. W. Holleman). National Technical Information Service, Springfield, Virginia, p. 91.

Damjanov, I., Solter, D. (1974a). Experimental teratoma. Curr. Top. Path. 59, 69.

Damjanov, I., Solter, D. (1974b). Host-related factors determine the outgrowth of teratocarcinomas from mouse eggcylinders. Z. Krebsforsch. Klin. Onkol. 81, 63.

Damjanov, I., Solter, D. (1974c). Embryo-derived teratocarcinoma elicit splenomegaly in syngeneic host. Nature (London) 249, 569.

Damjanov, I., Solter, D., Skreb, N. (1971). Teratocarcinogenesis as related to the age of embryos grafted under the kidney capsule. Wilhelm Roux' Arch. 167, 288.

Ferrer, J. F. (1968). Role of the spleen in passive immunological enhancement. Transplantation 6, 167.

Fidler, I. J. (1974). Immune stimulation-inhibition of experimental cancer metastasis. Cancer Res. 34, 491.

Gooding, L. R., Edidin, M. (1974). Cell surface antigens of a mouse testicular teratoma. Identification of an antigen physically associated with H-2 antigens on tumor cells. J. Exp. Med. 140, 61.

Markert, C. L. (1968). Neoplasia: a disease of cell differentiation. Cancer Res. 28, 1908.

McKhann, C. F., Jagarlamoody, S. M. (1971). Evidence for immune reactivity against neoplasms. Transplantation Rev. 7, 55.

Meier, H., Myers, D. D., Fox, R. R., Laird, C. W. (1970). Occurrence, pathological features, and propagation of gonadal teratomas in inbred mice and in rabbits. Cancer Res. 30, 30.

Pierce, G. B., Jr. (1967). Teratocarcinoma: model for a developmental concept of cancer. Curr. Top. Develop. Biol. 2, 223.

Prehn, R. T. (1972). The immune reaction as a stimulator of tumor growth. Science 176, 170.

Prehn, R. T., Lappe, M. A. (1971). An immunostimulation theory of tumor development. Transplantation Rev. 7, 26.

Skreb, N., Damjanov, I., Solter, D. (1972). Teratomas and teratocarcinomas derived from rodent egg-shields. In "Cell Differentiation" (Eds. R. Harris, P. Alin and D. Viza). Munksgaard, Copenhagen, p. 151.

Slye, M., Holmes, H. S., Wels, H. G. (1919). Primary spontaneous tumors of the testicle and seminal vesicle in mice and other animals. J. Cancer Res. 4, 207.

Solter, D., Damjanov, I., Koprowski, H. (1975). Embryo-derived teratoma: a model system in developmental and tumor biology. In "The Early Development of Mammals" (Eds. M. Balls and A. E. Wild). Cambridge University Press, London, p. 243.

Solter, D., Skreb, N., Damjanov, I. (1970). Extrauterine growth of mouse egg-cylinders results in malignant teratoma. Nature (London) 227, 503.

Stevens, L. C. (1964). Experimental production of testicular teratomas in mice. Proc. Nat. Acad. Sci. U.S. 52, 654.

Stevens, L. C. (1966). Development of resistance to teratocarcinogenesis by primordial germ cells in mice. J. Nat. Cancer Inst. 37, 859.

Stevens, L. C. (1967). The biology of teratomas. Advan. Morph. 6, 1.

Stevens, L. C. (1968). The development of teratomas from intratesticular grafts of tubal mouse eggs. J. Embryol. Exp. Morph. 20, 329.

Stevens, L. C. (1970a). Experimental production of testicular teratomas in mice of strains 129, A/He, and their F_1 hybrids. J. Nat. Cancer Inst. 44, 929.

Stevens, L. C. (1970b). Environmental influences on experimental teratocarcinogenesis in testes of mice. J. Exp. Zool. 174, 407.

Stevens, L. C. (1970c). The development of transplantable teratocarcinomas from intratesticular grafts of pre- and postimplantation mouse embryos. Develop. Biol. 21, 364.

Stevens, L. C. (1973). A new inbred subline of mice (129/ter Sv) with a high incidence of spontaneous congenital testicular teratomas. J. Nat. Cancer Inst. 50, 235.

Stevens, L. C., Little, C. C. (1954). Spontaneous testicular tumors in an inbred strain of mice. Proc. Nat. Acad. Sci. U.S. 40, 1080.

Stevens, L. C., Mackensen, J. A. (1961). Genetic and environmental influences on teratocarcinogenesis in mice. J. Nat. Cancer Inst. 27, 443.

Stevens, L. C., Varnum, D. S. (1974). The development of teratomas from parthenogenetically activated ovarian mouse eggs. Develop. Biol. 37, 369.

MUTAGENIZED CLONES OF A PLURIPOTENT TERATOMA CELL LINE: VARIANTS WITH DECREASED DIFFERENTIATION OR TUMOUR-FORMATION ABILITY

THIERRY BOON[1], ODILE KELLERMANN,
ELISABETH MATHY AND JEAN A. GAILLARD

Laboratoire de Genetique Cellulaire
de l'Institut Pasteur
et du College de France

INTRODUCTION

Our studies on cell determination are based on the use of pluripotent teratoma cell lines (Rosenthal et al., 1970; Kahan et al., 1970; Jakob et al., 1973). When these permanent clonal lines are injected into syngeneic mice, they produce progressive tumours containing a variety of differentiated tissues of ectodermal, endodermal and mesodermal origin. We thought that it would be useful to have a set of cell variants having acquired specific restrictions of their differentiation potential. We therefore undertook to analyze *in vivo* the differentiation of a set of clones isolated from a population of a pluripotent teratoma line previously treated with the mutagen nitrosoguanidine. In the course of these experiments, we have found that such a mutagenized population not only contains differentiation variants but also variants whose ability to produce progressive tumours in syngeneic mice is severely reduced.

The variants were isolated from the pluripotent teratoma line PCC4.aza1. This is a mutant resistant to 8-azaguanine

---

[1]Present address: Institute of Cellular and Molecular Pathology, Avenue Hippocrate, 75, B-1200 Bruxelles, Belgium, Tel. (02) 762.34.00.

isolated from line PCC4 (Jakob et al., 1973). The injection of PCC4.azal cells into syngeneic 129 mice results in the appearance within a few weeks of large progressive tumours containing differentiated cell types derived from all three germ layers. Like PCC4, PCC4.azal does not differentiate *in vitro*. Its karyotype is pseudodiploid: it has a mode at 39 chromosomes, including two metacentrics (Guenet et al., 1974).

EXPERIMENTAL PROCEDURE

Most of the procedures used have been described previously (Jakob et al., 1973). Mutagenesis was carried out with N-methyl-N-nitro-N-nitrosoguanidine at a concentration of 3 µg/ml in Earle balanced salt solution. The mutagenized cells were allowed to multiply *in vitro* for three days in order to allow the mutations to segregate. Cloning was then performed by distributing a dilute suspension of single cells in a number of small wells (Jakob et al., 1973) such that less than 15% of the clones could have been derived from more than one cell. After each clone had grown, it was injected at a dose of 5.10^5 cells into 129 mice and aliquots were frozen for further analysis. When tumours were obtained, they were analyzed histologically.

A control population was treated exactly in the same way, except that nitrosoguanidine was not added. Ten control clones were isolated. They all produced differentiated tumours like PCC4.azal.

We have performed three mutagenesis experiments. The fraction of cells able to multiply *in vitro* after the nitrosoguanidine treatment was on the order of 0.1 to 1%.

In one of these experiments, we controlled the efficiency of the mutagenesis by measuring the frequency of mutants able to multiply in a medium containing 1 µg/ml amanitin. We found that the frequency of these mutants was about 10^{-5} in the mutagenized population. This represents at least a forty-fold increase over the level found in the unmutagenized control.

ABBREVIATIONS AND TERMINOLOGY

dif⁻ - clones showing a reduced ability to differentiate *in vivo*.
tum⁻ - clones showing reduced tumorigenicity in strain 129 mice.

RESULTS AND CONCLUSIONS

We have obtained a few clones that have a significantly reduced ability to differentiate *in vivo*; we call them dif^-. We have also obtained a number of clones, whose ability to produce tumours in 129 mice is very much reduced. We call them tum^-. The results can be summarized as follows:
Experiment A: 15 clones were analyzed. We obtained 2 dif^- and 7 tum^-.
Experiment B: 12 clones we analyzed. We obtained 1 dif^- and 3 tum^-.
Experiment C: 38 clones were analyzed. We obtained 2 dif^- and 2 tum^-.
The numbers of variants are even higher than those indicated here, as we have chosen not to call dif^- or tum^- those clones which show only a slightly decreased potential for differentiation or tumour formation.

Analysis of dif^- Clones

Our initial hope was to obtain dif^- variants with specific restrictions in their differentiation potential. However, all our variants show a general reduction of their ability to differentiate. Furthermore, their differentiation defect is not an absolute one; an occasional tumour does contain ectodermal, endodermal and mesodermal derivatives, although usually in reduced amount.

As an example, we give here the results obtained with two rather extensively analyzed dif^- variants:
Clone 35: of 22 tumours, 18 were completely undifferentiated, 2 contained derivatives from one or two germ layers and 2 contained derivatives of the three germ layers.
Clone 39: of 27 tumours, 23 were completely undifferentiated, 2 contained a very small amount of ectodermal tissue (primitive neural tissue) and 2 contained derivatives of the three germ layers.
As a reference, of 23 PCC4.azal tumours analyzed, none was undifferentiated, 2 contained derivatives of only one or two germ layers and 21 contained ectodermal, mesodermal and endodermal derivatives.

The karyotypes of clones 35 and 39 are hypotetraploid: most metaphases contain between 69 and 75 chromosomes. This may be the cause of their dif^- phenotype. Possibly, the occasional differentiated tumours originate from pseudodiploid segregants.

It seemed worthwhile to test for the presence of the F9

antigen on clones 35 and 39, as this antigen is only found on 'primitive' teratocarcinoma lines, morulae and sperm (Artzt et al., 1973). Cytotoxic tests performed with these clones showed them to be about as sensitive to anti-F9 antibodies as PCC4.aza1.

Analysis of tum^- Clones

A number of clones have a reduced ability to produce tumours in syngeneic mice after subcutaneous or intraperitoneal injection. Usually, no tumour appears. Sometimes, a tumour appears at about the normal time but it either regresses or stops increasing in size.

The results of injection of a number of tum^- clones is described in the second column of Table 1. Clearly, the tumourigenicity of these clones is very much reduced compared to that of PCC4.aza1. It should also be noted that the rare

TABLE 1

RESULTS OF INJECTIONS OF TUM^- CLONES
INTO SYNGENEIC STRAIN 129 MICE

Clone	Unirradiated Mice		Irradiated Mice	
PCC4.aza1 (control)	26/26	(29 days)	6/6	(28 days)
20	2/20	(51 days)	6/6	(18 days)
21	3/15	(74 days)	5/5	(50 days)
25	0/20		6/6	(22 days)
33	3/36	(49 days)	10/11	(25 days)
40	2/16	(35 days)	6/6	(31 days)
42	0/8		3/3	(29 days)
43	0/15			
51	2/17	(41 days)	6/6	(26 days)
70	0/9		3/3	(21 days)
133	0/6		2/2	(25 days)

This table indicates the number of animals that have a progressive tumour within 60 days/total number of animals injected. The average number of days required to obtain a tumour with a diameter of about 0.5 cm are noted in parentheses.

Strain 129 mice, about 70 days old, were injected with doses of cells varying between 5.10^5 and 2.10^6. The injection medium was Dulbecco modified Eagle's medium containing 1% fetal calf serum. The mice were either not irradiated or given 600 rads of γ irradiation from a cesium source immediately prior to injection.

tumours obtained with these clones reach a large size significantly later than the tumours produced by PCC4.aza1. The tum^- phenotype is stable; tum^- clones maintained in culture over more than 50 generations remain unable to make tumours.

The karyotype of some tum^- clones (e.g. 20, 25, 43) is nearly diploid like that of PCC4.aza1. Clones 21, 40 and 51 have a hypotetraploid karyotype. Clones 20, 25, 40 and 43 are about as sensitive as PCC4.aza1 to the cytotoxic anti-F9 antiserum.

It appeared worthwhile to examine why these variants are unable to produce tumours. A first possibility is that they simply multiply less well than PCC4.aza1. This is not observed *in vitro* for most of the variants. Still, these clones could have been modified so as to be less able to multiply in the metabolic and hormonal environment of the mouse. This hypothesis is rendered unlikely by the results of irradiation experiments shown in Table 1. When the mice are irradiated with 600 rads prior to the injection of the cells, the tum^- clones produce progressive tumours almost every time. Moreover, with the exception of clone 21, these tumours grow as fast as the PCC4.aza1 tumours.

It is generally accepted that irradiated mice have a decreased ability to reject allogeneic and syngeneic tumour cells. It appears, therefore, that the tum^- phenotype is due to an increased vulnerability to the defences of the host, whether these are specific immune response processes or less specific processes.

The loss of tumourigenicity of the tum^- variants and that observed with differentiated teratoma-derived lines, like PCD1 (Boon et al., 1974), probably do not have the same origin. Line PCD1, which has muscle-like properties does not produce tumours, even when injected into irradiated mice (Boon et al., 1974). *In vitro*, the cells show contact inhibition. On the other hand, most tum^- variants (e.g. clones 20 and 25) are not contact inhibited. Moreover, the analysis of the tumours obtained in irradiated animals demonstrates that many of these variants (e.g. 20 and 25) have remained pluripotent.

We believe that the tum^- variants and their tum^+ control, PCC4.aza1, offer a particularly favorable system for the study of the interaction between a cancerous cell and the rejection mechanisms of the host. Indeed, as far as we know, there are no other systems where two closely related cells, both potentially able to multiply *in vivo*, appear to differ only by the way they stimulate the host defences or by their sensitivity to these defences. We shall study this system with tests that have been devised to evaluate the cellular immune response *in vitro*. Any test relevant to the *in vivo* situation ought to show a difference between tum^- variants and their tum^+ control.

ACKNOWLEDGEMENTS

This work was supported by grants from the "Centre National de la Recherche Scientifique," the National Institutes of Health and the Andre Meyer Foundation.

REFERENCES

Artzt, K., Dubois, P., Bennett, D., Condamine, H., Babinet, C., Jacob, F. (1973). Surface antigens common to mouse cleavage embryos and primitive teratocarcinoma cells in culture. Proc. Nat. Acad. Sci. U.S. 70, 2988.

Boon, T., Buckingham, M. E., Dexter, D. L., Jakob, H., Jacob, F. (1974). Teratocarcinome de la souris: isolement et proprietes de deux lignees de myoblastes. Ann. Microbiol. (Inst. Pasteur) 125B, 13.

Guenet, J. L., Jakob, H., Nicolas, J. F., Jacob, F. (1974). Teratocarcinome de la souris: etude cytogenetique des cellules a potentialities multiples. Ann. Microbiol. (Inst. Pasteur) 125A, 135.

Jakob, H., Boon, T., Gaillard, J., Nicolas, J. F., Jacob, F. (1973). Teratocarcinome de la souris: isolement, culture et proprietes de cellules a potentialites multiples. Ann. Microbiol. (Inst. Pasteur) 124B, 269.

Kahan, B. W., Ephrussi, B. (1970). Developmental potentialities of clonal *in vitro* cultures of mouse testicular teratoma. J. Nat. Cancer Inst. 44, 1015.

Rosenthal, M. D., Wishnow, R. M., Sato, G. H. (1970). *In vitro* growth and differentiation of clonal populations of multipotential mouse cells derived from a transplantable testicular teratocarcinoma. J. Nat. Cancer Inst. 44, 1001.

V.
Control of Differentiation of Embryonal Carcinoma Cells

THE FORMATION OF EMBRYOID BODIES *IN VITRO* BY HOMOGENEOUS EMBRYONAL CARCINOMA CELL CULTURES DERIVED FROM ISOLATED SINGLE CELLS

GAIL R. MARTIN[1] AND MARTIN J. EVANS

Dept. of Anatomy and Embryology
University College London
London WC1E 6BT, England

INTRODUCTION

Embryonal carcinoma cells, the stem cells of teratocarcinomas, are a useful alternative to early embryos for the study of processes of mammalian cell determination and differentiation because of their close similarity to the pluripotent cells of the early embryo. The similarities between teratocarcinoma stem cells and early embryo cells are fully examined in this volume, and have also been reviewed recently by Damjanov and Solter (1974) and by Martin (1975).

The advantages of studying the behavior of these cells *in vitro* are obvious. In order for the full potential of the tissue culture system to be realized, however, it must be possible to isolate *clonal* pluripotent embryonal carcinoma cell lines which will grow in the undifferentiated state and which will also differentiate to a wide variety of cell types under defined conditions *in vitro*. The reason that emphasis is placed on the isolation of clonal cell lines is that only by using cultures which originate from a single cell can one be certain that when differentiation of the cells does occur, it is cell determination (the process by which multipotential cells become committed to a particular developmental pathway)

---

[1]Present address: Dept. of Pediatrics, University of California, San Francisco, California 94143.

as well as subsequent terminal differentiation that is being observed. Both of these problems--the growth of clonal embryonal carcinoma cells in the undifferentiated state and the conditions under which they will differentiate *in vitro*--are the subject of this paper and the one which follows (Evans and Martin, 1975).

The teratocarcinomas from which pluripotent embryonal carcinoma cells can be isolated occur as solid tumors, which can sometimes be converted to growth in an ascitic form. The solid growths usually contain a wide variety of differentiated cell types, as well as embryonal carcinoma cells which are found haphazardly distributed throughout the tumor in the form of small groups or nests. When the cells grow in suspension in the peritoneal cavity, they form aggregates which are called embryoid bodies because of their resemblance to certain stages of embryogenesis (Stevens, 1959, 1960; Pierce, Dixon and Verney, 1960). Two types of embryoid body can be obtained, simple and cystic. Simple embryoid bodies consist of an inner core of embryonal carcinoma cells surrounded by a single layer of endodermal cells. They resemble the embryonic portion of the 5-day mouse embryo, although they are often incorrectly referred to as morula-like structures. At the morula stage, the pluripotent cells of the embryo are surrounded not by endoderm, but by a single layer of prospective trophoblast (reviewed by Herbert and Graham, 1974). Cystic embryoid bodies are more complex, many of them containing embryonal carcinoma cells, a variety of differentiated tissues and a fluid-filled cyst. These structures bear striking similarities to the fetal portion of older mouse embryos, but are clearly disorganized in comparison with them (Hsu and Baskar, 1974).

A number of clonal lines of teratocarcinoma stem cells have been isolated from tumors in strain 129 mice. Since embryonal carcinoma cells form a large proportion of the cells in embryoid bodies, it is an advantage to begin with these structures. Several groups have reported the isolation of clonal embryonal carcinoma cell lines from embryoid bodies (Kahan and Ephrussi, 1970; Rosenthal et al., 1970; Jakob et al., 1973; Bernstine et al., 1973). It is also possible to obtain embryonal carcinoma cell lines from the solid form of the tumor (Evans, 1972; Martin and Evans, 1974, 1975; Jami and Ritz, 1974), although this is more difficult if the tumor is well-differentiated and contains relatively few stem cells.

The pluripotent clonal cell lines which have been isolated from teratocarcinomas fall into two categories: homogeneous embryonal carcinoma cell cultures and heterogeneous cultures which contain embryonal carcinoma cells. The pluripotency of these cell lines has been demonstrated by injecting the cells into syngeneic hosts and determining that the tumors formed contain derivatives of all three primary germ layers. Since most subclones of pluripotent clonal cell lines can also give rise to multidifferentiated teratocarcinomas, it is clear that the pluripotency of the cells can be maintained during growth *in vitro*.

The various homogeneous cell lines which have been described have different growth requirements. Some have been found to be dependent on the addition of feeder cells (Kahan and Ephrussi, 1970; Martin and Evans, 1975), while others are feeder-independent in their growth, although some of these cell lines require gelatin-coated substrata (Jakob et al., 1973; Bernstine et al., 1973; Jami and Ritz, 1974). The heterogeneous clonal teratocarcinoma cell lines do not require added feeder cells, and this is probably because, in addition to embryonal carcinoma cells, they already contain fibroblastic or epithelioid cells which probably serve as feeder cells (Rosenthal et al., 1970; Evans, 1972; Martin and Evans, 1974).

Recent results (Martin and Evans, 1975) have helped to clarify differences among the various embryonal carcinoma cell lines, and have enabled us to make certain generalizations about the growth and differentiation of embryonal carcinoma cells *in vitro*.

MATERIALS AND METHODS

All cells were cultured in Dulbecco's modified Eagle's medium supplemented with 10% calf serum (selected batches) and all clonal cell lines were found to be free of mycoplasma infection as determined by the ^{3}H-thymidine-autoradiography method (Studzinski et al., 1973).

Primary and Secondary Cultures from Teratocarcinomas

Primary cultures can be prepared from solid tumors in the following way: the tumors are removed from the animal, washed with phosphate-buffered saline and finely minced with a scissors. The small pieces of tumor are disaggregated by agitating them in a solution of 1 mg/ml collagenase (Sigma, Type 1) in tris-buffered saline. The cell suspension is decanted and centrifuged and the remaining pieces of tumor are agitated with a second volume of collagenase solution. The cells from the first and second disaggregation are seeded on feeder layers at approximately 5×10^6 cells/9 cm dish.

The cultures are incubated at 37° and when they become dense they are disaggregated and secondary cultures are prepared by seeding the cells on STO (mouse fibroblast) feeder layers at approximately 5×10^6 cells/9 cm dish. The nature and preparation of the feeder layers have been detailed elsewhere (Martin and Evans, 1975).

If embryoid bodies are used as the starting material, they are washed with phosphate-buffered saline, allowed to settle in a conical centrifuge tube, resuspended in fresh medium and seeded in tissue culture dishes. Those which do not attach within 24 hours can be removed from the petri dish, washed and reseeded in fresh medium in another tissue culture dish. After several days, these primary cultures are disaggregated and secondaries are prepared by seeding the cells at 5×10^6 cells/9 cm dish containing feeder layers.

Growth of Clonal Embryonal Carcinoma Cell Lines

In order to obtain single embryonal carcinoma cells from which to establish clonal cell lines, one can begin with either the solid form of the tumor or embryoid bodies. If the latter are used, or if the solid tumor contains a high proportion of embryonal carcinoma cells, then cloning is usually successful from disaggregated primary or secondary cultures of the tumor. If one begins with a well-differentiated solid tumor containing only a small number of embryonal carcinoma cells, it is usually best to select areas of the primary or secondary cultures which clearly contain embryonal carcinoma cells, to disaggregate only those areas and to select single cells from them.

There are obvious advantages in initiating clonal cultures from isolated single cells rather than in using techniques which depend on the statistical probability of beginning with a single cell (such as colony isolation or the dilution method of cloning). Jami and Ritz (1974) first established clonal embryonal carcinoma cell lines using single cell isolation techniques. The method we have used to initiate clonal embryonal carcinoma cell lines is a modification of the method of Macpherson (1973) and is described elsewhere (Martin and Evans, 1975). The embryonal carcinoma cell lines we have obtained from isolated single cells are listed in Table 1. Stock cultures of pluripotent clonal cell lines are maintained in the undifferentiated state by passaging the cells every three days. The cultures are disaggregated by trypsinization and the cells are seeded at 5×10^6 cells/9 cm dish containing a feeder layer. The medium is changed on the second day after subculturing. Approximately 5×10^7 cells

TABLE 1

TUMORS USED FOR DERIVATION OF EMBRYONAL CARCINOMA CELL CULTURES

	Animal	Genotype	Tumor of Origin	Source, Sex of Tumor	Cell Line Name
Type I.	mouse	129	LS 402C-1684[a]	spontaneous, male	Nulli-SCC1
	"	"	"	"	Nulli-SCC2
Type II.	mouse	129SvSl^{J}CP	OTT-5568[b]	3d embryo	SCC-S1
	"	"	"	"	SCC-S2
	"	"	"	"	SCC-S3
	"	"	"	"	SCC-S4
Type III.	mouse	129SvSl^{J}CP	OTT-5568	3d embryo	SCC-PSA2
	"	"	"	"	SCC-PSA4
	"	"	"	"	SCC-PSA5
	"	"	"	"	OSB

[a] Stevens, 1958
[b] Stevens, 1970

are harvested at each passage. In order to obtain differentiation of the cells, cultures are seeded at 10^7 cells/9 cm tissue culture dish without added feeder cells.

Stock cultures of nullipotent cells can be maintained on feeder layers as are the pluripotent cell lines, but they can also be maintained in the absence of feeder cells on gelatin-coated dishes. These are prepared by incubating tissue culture dishes with a 0.1% solution of gelatin (Swine Skin, Type 1, Sigma, selected batches), for two hours or more at room temperature. The gelatin is then removed and the cells are seeded at 6.5×10^6 cells/9 cm dish. The cells are subcultured every three days, and approximately 3.5×10^7 cells are harvested at each passage. In order to obtain nullipotent aggregates, the cells are seeded at 5×10^6 cells/9 cm tissue culture dish (no gelatin coating).

RESULTS

Feeder Dependence of Embryonal Carcinoma Growth

In our experience, embryonal carcinoma cells freshly isolated from the tumor cannot grow *in vitro* without the addition of feeder cells. Although other workers have clearly been able to establish feeder-independent clonal embryonal carcinoma cell lines, we believe that this involves a certain amount of selection and adaptation to tissue culture which can affect the ability of the cells to differentiate *in vitro*, and we have therefore avoided such procedures.

There is nothing unusual in the observation that embryonal carcinoma cells require the presence of feeder cells for growth at low density; the same is true for many other cell types. We have found, however, as previously noted by Kahan and Ephrussi (1970), that even at very high density the growth of embryonal carcinoma cells is dependent on the presence of feeder cells. When the cells are subcultured in the absence of added feeder cells the cultures are homogeneous, and the cells have the typical morphology of embryonal carcinoma cells; few, if any, of the feeder cells persist after subculture (Fig. 1). Such homogeneous embryonal carcinoma cells grow well in the absence of added feeders for one or two serial passages, albeit more slowly than in the presence of feeder cells, but in subsequent serial passages the cells die. Cells which would otherwise die in the absence of feeder cells can be rescued by plating on feeder layers, and these rescued cells will multiply at the same rate as cells which have never been cultured in the absence of feeders (Fig. 2).

These data are illustrated in Figure 2.

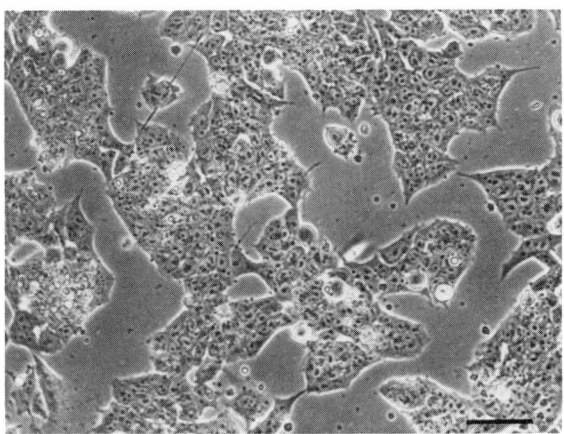

Figure 1. Homogeneous population of pluripotent embryonal carcinoma cells derived from an isolated single cell. Clone SCC-S2 thirty-six hours after the second passage in the absence of feeder cells. Phase contrast microscopy. Scale bar = 120 um.

Since the growth of some feeder-independent clonal embryonal carcinoma cell lines requires the presence of a gelatin-coated substratum (Jakob et al., 1973; Bernstine et al., 1973), we have attempted to substitute gelatin-coating for feeder layers. While the gelatin appears to increase the spreading and attachment of the cells to the substratum, and presumably also increases the survival of the cells, feeder cells must have some additional effect since the cells do not grow well in their absence, even on gelatin-coated dishes (Fig. 2c). Preliminary experiments suggest that feeder-conditioned medium also does not substitute for feeder cells.

Early Stages in the Differentiation of Clonal Embryonal Carcinoma Cells *In Vitro*

Our observations have led us to the conclusion that, except in certain circumstances (see Evans and Martin, 1975), embryonal carcinoma cells do not differentiate *in vitro* in the presence of feeder cells. However, if one examines the cultures in the second passage in the absence of feeder cells, when the cells are still growing well, one observes certain morphological changes occurring: initially the cells are homogeneous embryonal carcinoma cells, which are found in groups that are well-attached to the substratum. Such clumps either detach from the surface of the dish spontaneously, or can be easily detached by gently pipetting medium over them. At the time when the clumps become fully three-dimensional, a morphological change is noticeable: the cells on the outer surface of the clumps form a distinct cell layer (Fig. 3a). These clumps now bear a striking resemblance to simple embry-

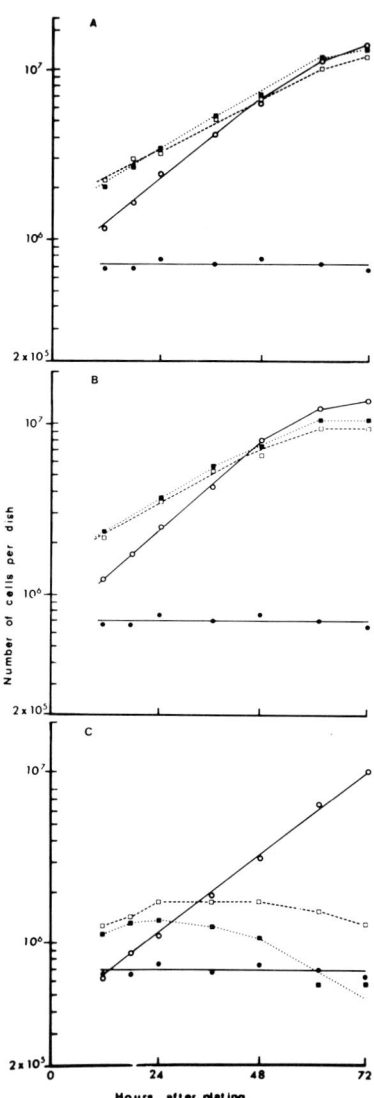

oid bodies found in the ascitic fluid of animals bearing intraperitoneal teratocarcinomas (Teresky et al., 1974). Results previously reported (Martin and Evans, 1975) and summarized below, confirm that these clumps are embryoid bodies.

In cross section, the clumps formed *in vitro* consist of a core of embryonal carcinoma cells separated from the outer cell layer by an eosinophilic extracellular material which stains with periodic acid Schiff, as does Reichert's membrane (Pierce et al., 1962) which is found in a similar location in embryoid bodies formed *in vivo*. The cells of the outer layer themselves stain with periodic acid Schiff, as would be expected if they were synthesizing this basement membrane in large quantities. Electron microscopy of the clumps confirmed that the inner cells have the distinctive structure of embryonal carcinoma cells; that is, a large nucleus and relatively little cytoplasm which is almost devoid of organelles except for mitochondria and a large number of free ribosomes. In contrast, the cells of the outer layer have the characteristics of the endodermal cells of embryoid bodies (Teresky et al., 1974). They contain plentiful swollen endoplasmic reticulum and have prominent microvilli on their outermost surface (Figs. 4, 5). In addition, high levels of the enzyme alkaline phosphatase are found in the cells of the inner core and very low levels in the cells of the outer layer (Fig. 6), as has been shown for embryoid bodies formed *in vivo* (Bernstine et al., 1973). Thus, by all criteria presently available, the clumps formed *in vitro* by homogeneous pluripotent clonal embryonal carcinoma cell lines derived from isolated single cells are identical to embryoid

---

Figure 2. Effect of feeder cells on the growth of pluripotent embryonal carcinoma cells. Stock cultures of clonal cell line SCC-S2 were plated at 2.5 x 10^6 cells/5 cm tissue culture dish (··■··), at 2.5 x 10^6 cells/5 cm tissue culture dish coated with gelatin (--▫--), or at 1.25 x 10^6 cells/5 cm tissue culture dish previously seeded with 1.25 x 10^6 Mitomycin-C treated STO feeder cells (——O——). At various times after plating, the number of cells/dish was determined, using a Coulter Counter. Each point on the curves represents an average of three determinations. For embryonal carcinoma cells plated on feeder layers, the curves shown represent the total number of cells present minus the number of feeder cells present on control feeder-only plates (——●——). Stock cells, previously passaged on feeder layers, were passaged in the absence of feeder layers (a) 0, (b) 1, or (c) 2 times prior to determination of the growth curves.

bodies formed *in vivo*.

Methods of Forming Embryoid Bodies *In Vitro*

The results described above have indicated one way in which embryoid bodies can be formed *in vitro*: homogeneous clonal embryonal carcinoma cells plated in the absence of feeder cells gradually form rounded clumps. When these become fully three-dimensional, by spontaneous or mechanical detachment from the substratum, an outer layer of endodermal cells becomes apparent. Embryoid bodies can also be formed *in vitro* in two other ways. Homogeneous embryonal carcinoma cells can be disaggregated by trypsinization and seeded on a non-adhesive substratum (for example, a bacteriological dish). The cells aggregate in suspension and between 12-36 hours after seeding, an endodermal cell layer becomes apparent on the outer surface of the aggregates. Embryoid bodies can also be formed *in vitro* by seeding embryonal carcinoma cells at very low densities on a feeder layer (for example 100 cells per dish), and then carefully removing the small colonies which form and seeding them on a non-adhesive substratum. The outer layer of endoderm appears within 12-36 hours of

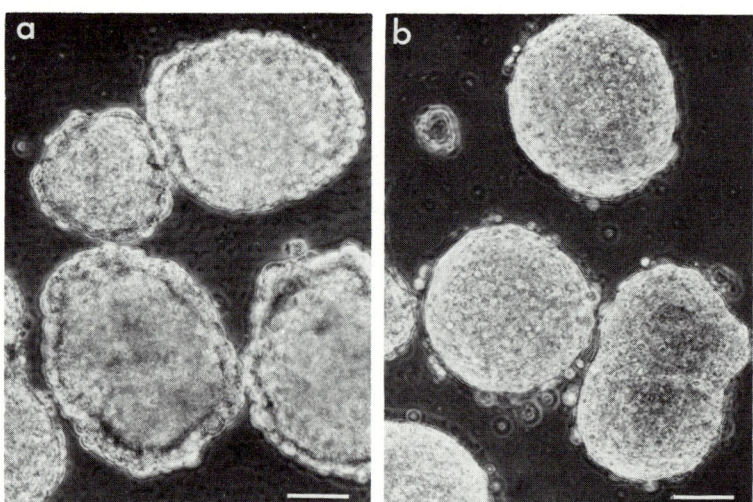

Figure 3. Clumps of embryonal carcinoma cells five days after plating a single cell suspension. The cell clumps were detached three days after plating. Phase contrast microscopy. Scale bar = 120 um. a, Pluripotent cells. The endodermal cell layer is apparent; b, nullipotent cells. No endodermal layer has formed.

placing these colonies in suspension. Using this method there is a high statistical probability that each of the embryoid bodies formed is itself a clone.

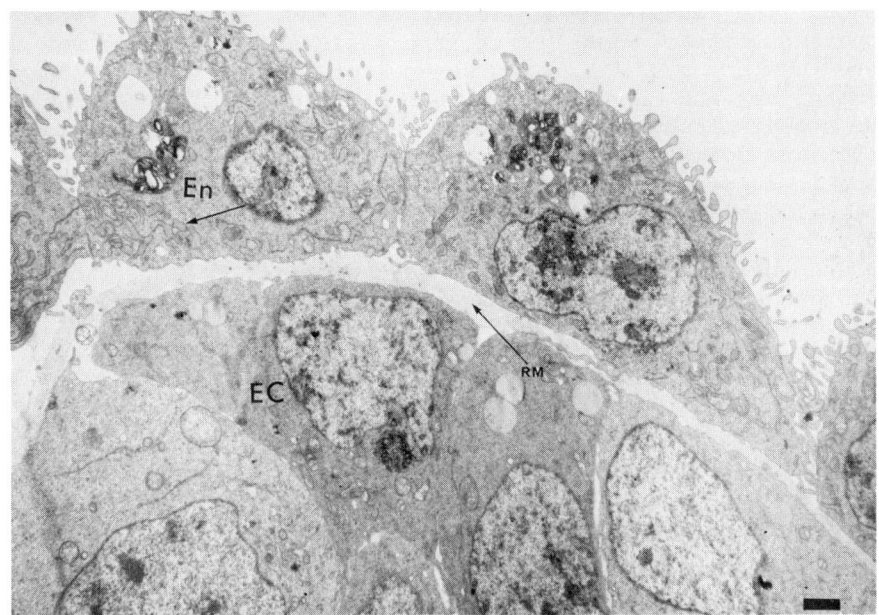

Figure 4. Electron micrograph showing part of an embryoid body formed in vitro. Scale bar = 1 um. EC: embryonal carcinoma cells in the core. En: endodermal cells of the outer layer, with plentiful swollen endoplasmic reticulum and prominent microvilli on their outermost surface. Note contrast with cytoplasm of embryonal carcinoma cells, which has few organelles. Arrow points to Reichert's membrane-like material, found between the inner and outer cells. From Martin and Evans, 1975.

Differences among Clonal Embryonal Carcinoma Cell Lines

The results described above were obtained with cultures established from single embryonal carcinoma cells derived from a pluripotent, clonal, heterogeneous teratocarcinoma cell line, SIKR (Evans, 1972; Martin and Evans, 1974). The embryoid bodies formed by these cells remain as simple embryoid bodies when kept in suspension, even over long periods of time. This is true for all four of the clones examined. Since the parent SIKR cells had been passaged for many generations *in vitro* before these clones were isolated, we applied the same single cell isolation techniques described above to obtain clonal lines derived from considerably earlier stocks of the parent SIKR line, and also from primary cultures of the tumor from which the SIKR cells were originally isolated. The results are basically the same, except for two important differences. First, the cells do not survive for more than one passage in the absence of feeder cells and they form clumps and then embryoid bodies in the first rather than the second passage off feeders. Second, the embryoid bodies which are formed by these clones do not remain simple when kept in suspension. Although the embryoid bodies formed initially by these clonal lines are virtually indistinguishable from those formed by the clones described above, if kept in suspension these embryoid bodies gradually undergo internal changes and become cystic (Fig. 7). This confirms the suggestion of Pierce and Dixon (1959) that cystic embryoid bodies arise from simple ones. Examination of histological sections of these cystic embryoid bodies formed *in vitro* indicates that

Figure 5. The outer surface of an embryoid body formed in vitro. *Scanning electron micrograph courtesy of Dr. A. Boyde. Scale bar = 10 um.*

they contain a variety of complex tissues and are identical to the cystic embryoid bodies found in some animals bearing intraperitoneal teratocarcinomas. The progression of these embryoid bodies from the simple form to the cystic form has many features in common with normal mouse embryogenesis, and this will be fully described elsewhere (Martin and Evans, in preparation).

In order to confirm that the formation of embryoid bodies is a consequence of the pluripotency of the cells, and in fact the first step in their differentiation, we examined the behavior of a nullipotent embryonal carcinoma cell line isolated in the same way as the pluripotent cells described above (Martin and Evans, 1975). These clones were prepared from

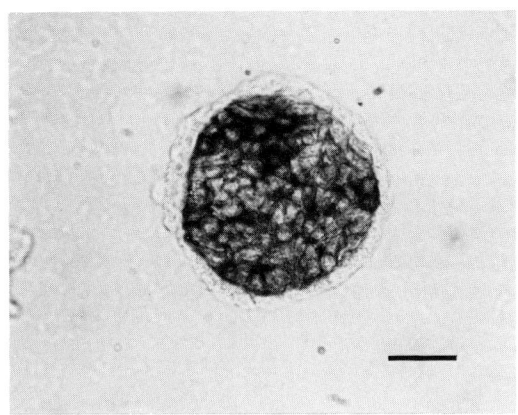

Figure 6. Localization of alkaline phosphatase activity in embryoid bodies formed by pluripotent cells in vitro. *Embryoid bodies were stained for the enzyme (Martin and Evans, 1975), embedded in plastic and sectioned. Embryonal carcinoma cells in the central core of the clump show a high enzyme activity. Endodermal cells of the outer layer show no activity. Scale bar = 60 um.*

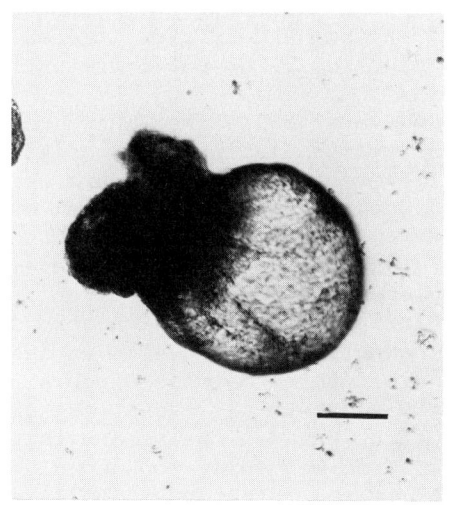

Figure 7. A cystic embryoid body formed in vitro. *Light microscopy of unfixed, unstained material. Scale bar = 36 um.*

primary cultures of a tumor which consists entirely of embryonal carcinoma; the cells of this tumor are apparently incapable of differentiation (Stevens, 1958). As expected, the cells behave in the same way as the pluripotent cells, in that they form clumps when plated in the absence of feeder cells, but unlike the pluripotent cells which then form an outer layer of endodermal cells, the nullipotent clumps remain completely homogeneous and never show any signs of differentiation *in vitro* (Fig. 3b). A second difference between the clonal nullipotent cell lines we have obtained, and all of the pluripotent cells described above is that the former are capable of growing for an indefinite number of serial passages in the absence of feeder cells, although they grow best when cultured on a gelatin-coated substratum.

DISCUSSION

We have shown that isolated single cells from a well differentiated teratocarcinoma can give rise to homogeneous cultures of embryonal carcinoma cells which can be maintained in the undifferentiated state by frequent subculture in the presence of feeder cells. We have also described the conditions under which these pluripotent embryo carcinoma cells will form simple embryoid bodies, the first stage in their differentiation *in vitro* (Martin and Evans, 1975; Evans and Martin, 1975). We have also obtained cultures from isolated single cells derived from an undifferentiated teratocarcinoma; these nullipotent cells do not differentiate *in vivo* or *in vitro*.

The clonal embryonal cell lines which we have isolated can be grouped into three distinct categories:

Type I. Nullipotent cells which form homogeneous aggregates; an outer cell layer of endodermal cells is never observed.

Type II. Pluripotent cells which form embryoid bodies which remain simple when kept in suspension.

Type III. Pluripotent cells which form simple embryoid bodies which subsequently become cystic in suspension.

We have described three methods of obtaining embryoid body formation *in vitro*. In all cases, the endodermal cell layer becomes apparent between twelve and thirty-six hours after the cells have formed fully rounded clumps or aggregates. This could be the result of a decrease in the growth rate of the cells, since a high rate of multiplication might be incompatible with a switch to differentiated functions. Our data indicate that when cells are seeded in the absence of feeder cells (method 1) they multiply more slowly (doubling time of 21 hours) than in the presence of feeder cells (doubling time of 13-15 hours). It seems likely, however, that the more important factor in the formation of endoderm is the change in the spatial configuration of the cells. This is supported by the observation that some clones (Type II) have the same growth rate in the first and second serial passages in the absence of feeders (Fig. 2), but these only form embryoid bodies in the second passage, when clump formation occurs.

It is possible that the appearance of the endoderm on the outer surface of aggregates or clumps of pluripotent embryonal carcinoma cells is due to the reassortment of an already heterogeneous population. This seems unlikely because the cultures were derived from isolated single cells, and endoderm formation occurs in cultures which are initially homogeneous. In order for reassortment to occur, one would have to postulate that a morphologically undetectable heterogeneity had arisen at some time during the cloning procedure or subsequent growth of the cells. It seems more likely that the embryonal carcinoma cells on the outside of the clumps differentiate to endoderm and the inner cells remain unchanged. This mechanism is similar to that proposed for the early events of embryogenesis in the mouse (Tarkowski and Wroblewska, 1967). The first determination that occurs in the mouse embryo is the formation of the trophectoderm. There is now good evidence (reviewed by Herbert and Graham, 1974) that the cells that become committed to form trophectoderm do so as a result of their position on the outer surface of the morula. In the development of the mouse embryo, the second cell type to appear after the establishment of the blastocyst, is the endoderm, which forms only on the free surface of the inner cell mass (adjacent to the blastocoel). The stimulus for the formation of endoderm may also be a positional one, as it has been found that when the inner cell mass is isolated from the trophectoderm, endoderm forms over the whole of its outer surface (Gardner and Papaioannou, 1975; Rossant, 1975).

Thus, the first event of differentiation of pluripotent mouse teratocarcinoma stem cells *in vitro* is similar to that of the inner cell mass of the mouse embryo. This suggests that in functional terms the embryonal carcinoma cells are equivalent to the inner cell mass cells of the 4-5 day blastocyst. The analogy between embryonal carcinoma cells and inner cell mass cells might seem imperfect because the endoderm formed by teratocarcinoma cells is not analogous to that which is formed by the cells of this inner cell mass: the latter does not produce Reichert's membrane-like material, which is produced by endodermal cells of embryoid bodies. Recently, however, Solter et al. (1974) have found that the proximal endoderm of blastocysts which have been removed from the animal and developed *in vitro* do produce Reichert's membrane-like material. The analogy between teratocarcinoma cell differentiation *in vitro* and mouse embryogenesis is further strengthened by our observations on the process of cystic embryoid body formation, which occurs when the simple embryoid bodies formed by some clonal embryonal carcinoma cultures (Type III) are kept in suspension.

All of the clonal embryonal carcinoma cell lines which we have isolated from a well-differentiated tumor, which was originally derived from a 3-day embryo, are pluripotent, as determined by reinjection of the cells into syngeneic mice and examination of the resultant tumors, and all of them behave in the same way: they form embryoid bodies under the conditions described above. The formation of embryoid bodies *in vitro* has also been observed in clonal cultures derived from a tumor originating from a 6-day embryo (Jami and Ritz, 1974) and from a spontaneous tumor originating from primordial germ cells (Rosenthal et al., 1970). However, in those experiments the significance of embryoid body formation was not appreciated as the primary event of differentiation of the cells since subsequent differentiation was poor, perhaps because these embryoid bodies were not allowed to attach to the substratum. Our results, described above, and also obtained with cell lines derived from teratocarcinomas in other strains of mice (Martin, manuscript in preparation), taken together with the observations of other authors, lead us to the conclusion that in general the normal differentiation of teratocarcinoma stem cells *in vitro* occurs via the formation of embryoid bodies. If some pluripotent embryonal carcinoma cell lines do not form embryoid bodies as the first step in their differentiation, we suggest that this may be because such cells have undergone either genetic or epigenetic changes, perhaps as part of their adaptation to tissue culture; such adaptation and selection is most likely to occur when the cells have been isolated in the absence of feeder cells.

The observations described here suggest that teratocarcinomas will be very useful in studying the early events of mouse embryogenesis, since the formation of the endodermal cell layer of embryoid bodies occurs under controlled conditions in large populations of cells. It is therefore now possible to use biochemical and immunological techniques (Stern, Martin and Evans, 1975) to study this particular cell determination and the determination and differentiation of other cell types which subsequently develop in the cultures.

ACKNOWLEDGEMENTS

We wish to thank Mr. R. Moss, Miss D. Bailey, Mrs. J. Astafiev, and Mr. A. Aldrich for their help in preparing the figures, and Miss A. Harns for typing the manuscripts, of both this paper and the one by Evans and Martin in this volume.

This investigation was supported by a grant from the Cancer Research Campaign. GRM is supported by a fellowship from the U.S/ Public Health Service, National Cancer Institute, 1 F22 CA04012-01.

REFERENCES

Bernstine, E. G., Hooper, M. L., Grandchamp, S., Ephrussi, B. (1973). Alkaline phosphatase activity in mouse teratoma. Proc. Nat. Acad. Sci., U.S. 70, 3899.
Damjanov, I., Solter, D. (1974). Experimental teratoma. Curr. Topics Path. 59, 69.
Evans, M. J. (1972). The isolation and properties of a clonal tissue culture strain of pluripotent mouse teratoma cells. J. Embryol. Exp. Morph. 28, 163.
Evans, M. J., Martin, G. R. (1975). The differentiation of clonal teratocarcinoma cell cultures *in vitro*. This volume, p. 237.
Gardner, R. C., Papaioannou, V. E. (1975). Differentiation in the trophectoderm and inner cell mass. In "The Early Development of Mammals" (Eds. M. Balls and A. Wild). Cambridge University Press, London, in press.
Herbert, M. C., Graham, C. F. (1974). Cell determination and biochemical differentiation of the early mammalian embryo. Curr. Topics Develop. Biol. 8, 151.
Hsu, Y. C., Baskar, J. (1974). Differentiation *in vitro* of normal mouse embryos and mouse embryonal carcinoma. J. Nat. Cancer Inst. 53, 177.
Jakob, H., Boon, T., Gaillard, J., Nicolas, J.-F., Jacob, F. (1973). Teratocarcinome de la souris: isolement, culture et proprietes de cellules a potentialites multiples. Ann. Microbiol. (Inst. Pasteur) 124B, 269.
Jami, J., Ritz, E. (1974). Multipotentiality of single cells of transplantable teratocarcinomas derived from mouse embryo grafts. J. Nat. Cancer Inst. 52, 1547.

Kahan, B. W., Ephrussi, B. (1970). Developmental potentialities of clonal *in vitro* cultures of mouse testicular teratomas. J. Nat. Cancer Inst. 44, 1015.

Macpherson, I. (1973). Animal cells. B. Microdrop techniques. In "Tissue Culture Methods and Applications" (Eds. P. F. Kruse and M. K. Patterson). Academic Press, New York, p. 241.

Martin, G. R. (1975). Teratocarcinomas as a model system for the study of embryogenesis and neoplasia: a review. Cell 5, 229.

Martin, G. R., Evans, M. J. (1974). The morphology and growth of a pluripotent teratocarcinoma cell line and its derivatives in tissue culture. Cell 2, 163.

Martin, G. R., Evans, M. J. (1975). The differentiation of clonal lines of teratocarcinoma cells: formation of embryoid bodies *in vitro*. Proc. Nat. Acad. Sci., U.S. 72, 1441.

Pierce, G. B., Jr., Dixon, F. J. (1959). Testicular teratomas. I. Demonstration of teratogenesis by metamorphosis of multipotential cells. Cancer 12, 573.

Pierce, G. B., Jr., Dixon, F. J. (1959). Testicular teratomas. II. Teratocarcinoma as an ascitic tumor. Cancer 12, 584.

Pierce, G. B., Jr., Dixon, F. J., Verney, E. L. (1960). Teratocarcinogenic and tissue forming potentialities of the cell types comprising neoplastic embryoid bodies. Lab. Invest. 9, 583.

Pierce, G. B., Jr., Midgley, A. R., Jr., Ram, J. S., Feldman, J. D. (1962). Parietal yolk sac carcinoma: clue to the histogenesis of Reichert's membrane of the mouse embryo. Amer. J. Path. 41, 549.

Rosenthal, M. D., Wishnow, R. M., Sato, G. H. (1970). *In vitro* growth and differentiation of clonal populations of multipotential mouse cells derived from a transplantable testicular teratocarcinoma. J. Nat. Cancer Inst. 44, 1001.

Rossant, J. (1975). Investigation of the determinative state of the mouse inner cell mass. II. The fate of isolated inner cell masses transferred to the oviduct. J. Embryol. Exp. Morph., in press.

Solter, D., Biczysko, W., Pienkowski, M., Koprowski, H. (1974). Ultrastructure of mouse egg cylinders developed *in vitro*. Anat. Rec. 180, 263.

Stern, P. L., Martin, G. R., Evans, M. J. (1975). Cell surface antigens of clonal teratocarcinoma cells at various stages of differentiation. Cell, in press.

Stevens, L. C. (1958). Studies on transplantable testicular teratomas of strain 129 mice. J. Nat. Cancer Inst. 20, 1257.

Stevens, L. C. (1959). Embryology of testicular teratomas in strain 129 mice. J. Nat. Cancer Inst. 23, 1249.

Stevens, L. C. (1960). Embryonic potency of embryoid bodies derived from a transplantable testicular teratoma of the mouse. Develop. Biol. 2, 285.

Stevens, L. C. (1970). The development of transplantable teratocarcinomas from intratesticular grafts of pre- and postimplantation mouse embryos. Develop. Biol. 21, 364.

Studzinski, G. P., Gierthy, J. F., Cholon, J. J. (1973). An autoradiographic screening test for mycoplasmal contamination of mammalian cell cultures. In Vitro 8, 466.

Tarkowski, A. K., Wroblewska, J. (1967). Development of blastomeres of mouse eggs isolated at the 4 and 8 cell stage. J. Embryol. Exp. Morph. 18, 155.

Teresky, A. K., Marsden, M., Kuff, E. L., Levine, A. J. (1974). Morphological criteria for the *in vitro* differentiation of embryoid bodies produced by a transplantable teratoma of mice. J. Cell Physiol. 84, 319.

DIFFERENTIATION OF TERATOMA CELL LINE PCC4.azal *IN VITRO*

MICHAEL I. SHERMAN

Roche Institute of Molecular Biology
Nutley, New Jersey   07110

The behavior of murine teratoma cells in culture has now been studied by a number of laboratories. Two kinds of clonal lines of teratoma embryonal carcinoma, or stem, cells have been characterized: in one case, the cells routinely give rise to one or more differentiated cell types *in vitro* (e.g., Rosenthal et al., 1970; Evans, 1972; Martin and Evans, 1974, 1975; Nicolas et al., 1975); on the other hand, a number of clonal lines of embryonal carcinoma cells have been found to be unable to differentiate *in vitro*, i.e., only a single characteristic cell morphology is observed (Fig. 1a). The latter class of clonal lines can themselves be divided into two groups on the basis of their behavior upon reinjection into isogenic mice: with some lines, the tumors are purely embryonal carcinoma, as would be expected from their *in vitro* behavior (Kahan and Levine, 1971; Bernstine et al., 1973; Artzt et al., 1973); in other instances, however, cells apparently unable to differentiate in culture can do so *in vivo* (Finch and Ephrussi, 1967; Kahan and Ephrussi, 1970; Jakob et al., 1973). Since these latter cell lines will remain as embryonal carcinoma under routine culture conditions, but do possess the capacity to differentiate, as indicated from tumor studies, they provide an ideal system for testing the ability of various effectors to trigger differentiation *in vitro*.

A number of years ago, Stevens (1958) noted that embryonal carcinoma cells resemble those of the preimplantation embryo. Since that time, a variety of different studies have supported this relationship (Stevens, 1968, 1970; Edidin et al., 1971; Artzt et al., 1973; Stevens and Varnum, 1974). It has recently become possible to study differentiation of preimplantation embryos *in vitro* (Hsu, 1971, 1973; Hsu et al., 1974; Sherman, 1972; Bell and Sherman, 1973; Sherman, 1975a). Because the cells of the inner cell mass (ICM) of the blastocyst share the multipotential properties of embryonal carcinoma cells, the pattern of differentiation of ICM cells *in vitro* will be described briefly for comparative purposes.

Hsu and coworkers (Hsu, 1971, 1973; Hsu et al., 1974) have demonstrated that a small percentage of mouse blastocysts cultured on a collagen substratum could continue to develop in a morphologically normal manner for several days. With the subsequent addition of human cord serum, ICM-derived structures strongly resembled ninth day embryos, complete with somites and beating heart. In simultaneous experiments carried out in this laboratory (Sherman, 1975a), a number of cell lines have been derived from blastocyst cultures. These cells bear a relationship to cells developing from teratomas *in vitro* and are described in detail elsewhere (see Sherman, 1975a, 1975c and Miller et al., 1975).

What is most germane to the experiments to be described here is a consideration of the early development of the ICM *in vitro*. In previous studies, it was indicated that shortly after attachment of the blastocyst to the culture dish, the ICM is exposed as a solid clump of cells adhering to the underlying trophoblast monolayer (Sherman, 1975a). At this time, ICM cells bear a striking morphological resemblance to embryonal carcinoma cells (Sherman, 1975c). Rather than dispersing and growing out along the culture dish with the trophoblast cells, the ICM clump continues to grow, but remains as a ball of cells. Only after a few days does this structure begin to visibly compartmentalize, often resembling an early egg cylinder, and only after about five or six days of culture do ICM derived cells begin to migrate from the central mass out along the culture dish. These migrating cells do not have the morphology of embryonal carcinoma cells; they are usually epithelioid or fibroblastic in appearance. Presumably, then, these cells have begun to

---

ABBREVIATIONS AND TERMINOLOGY

DME - Dulbecco-modified Eagle's medium.
ICM - inner cell mass of the blastocyst.

For the purposes of this report, cells with morphologies other than that of embryonal carcinoma cells will be referred to as 'differentiated.'

differentiate, under some unknown influence, while still within the central mass of ICM derived cells, or very shortly thereafter.

ICM's containing reduced numbers of cells in 'miniblastocysts' derived from disaggregated blastomeres usually fail to differentiate, or indeed survive, *in vitro* (Sherman, 1975b). A critical mass therefore appears to be required for successful development of ICM cells. By analogy, it might be possible that embryonal carcinoma cell lines failing to differentiate *in vitro* may not be capable of reaching a necessary critical mass. Experiments were carried out on one such cell line in order to test this hypothesis.

GROWTH OF PCC4.azal CELLS AT HIGH DENSITIES

The cells used in this study, PCC4.azal, were clonally derived from embryoid bodies, ultimately from teratocarcinoma OTT 6050 (Stevens, 1970), and then selected for resistance to high levels of 8-azaguanine (Jakob et al., 1973). Although they form characteristic multidifferentiated teratomas when injected into mice, the cells under normal tissue culture conditions maintain only a single morphology that is typical of embryonal carcinoma (Fig. 1a). As described by Jakob

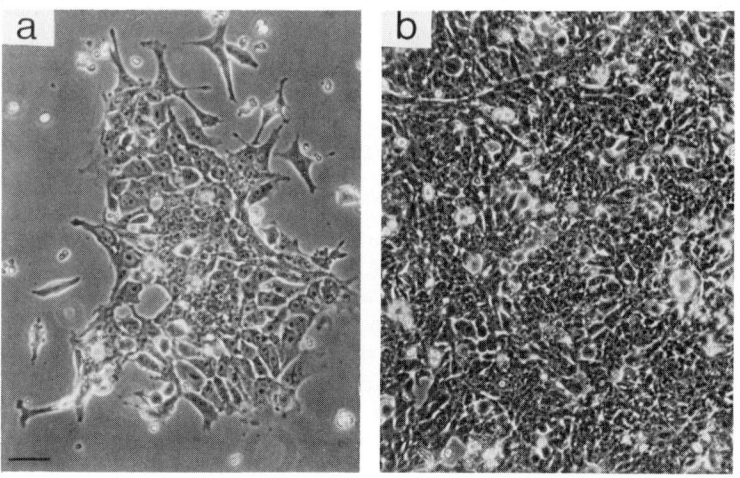

Figure 1. Morphology of PCC4.azal embryonal carcinoma cells (a) under normal culture conditions and (b) growing at high density. Scale marker = 50 um.

and coworkers (1973), the cells grow rapidly in Dulbecco-modified Eagle's (DME) medium, but they quickly degenerate and lyse at high densities. Initially, efforts were made to generate cells from this line which were capable of growing at high density, to see if this would be adequate to trigger differentiation. To accomplish this, cells were seeded at high density, and maintained with only infrequent medium changes (once weekly). This technique was successful in some cases (e.g., Fig. 1b). However, even though cultures were maintained in monolayers at high densities for several weeks, they did not produce cells which were morphologically different from embryonal carcinoma cells.

PRODUCTION OF AGGREGATES OF PCC4.azal CELLS

Since ICM cells differentiate from solid clumps, and since embryoid bodies are often multipotential when placed into culture (Rosenthal et al., 1970; Kahan and Ephrussi, 1970), it was reasoned that the two dimensional matrix of a monolayer might not be adequate to trigger differentiation of PCC4.azal cells. Consequently experiments were next carried out with solid cell aggregates of embryonal carcinoma cells. Rosenthal et al. (1970) have previously demonstrated that aggregates of teratoma cells can be generated by culture in bacterial petri dishes (which do not have the surface treated to allow cell adhesion). Some of these aggregates were, in fact, found to contain differentiated cell types. Kahan and Ephrussi (1970) made similar observations on aggregates formed by placing culture flasks on a gyratory shaker.

In this study, aggregates were initially generated by culturing cells on Nucleopore filters (Nucleopore Corporation, Pleasanton, California). These polycarbonate filters are opalescent or translucent, depending upon the pore size. When PCC4.azal cells are plated onto Nucleopore filters, they do not adhere, but proliferate into multicellular balls which, within a week, reach diameters up to 1-2 mm (Fig. 2a). When these aggregates are placed on a normal plastic tissue culture surface, the majority attach (Fig. 2b; Table 1, Expt. 1) and flatten out as a disc of cells with a dense core. By 24 hours, cells have begun to grown out from most of the clumps. When inspected seven days after attachment to the tissue culture surface, it was observed that almost 40% of the surviving clumps contained at least some outgrowing cells which did not share the morphology of embryonal carcinoma cells; characteristically, these cells were fibroblastic in appearance, and were flatter than, and lacked the promi-

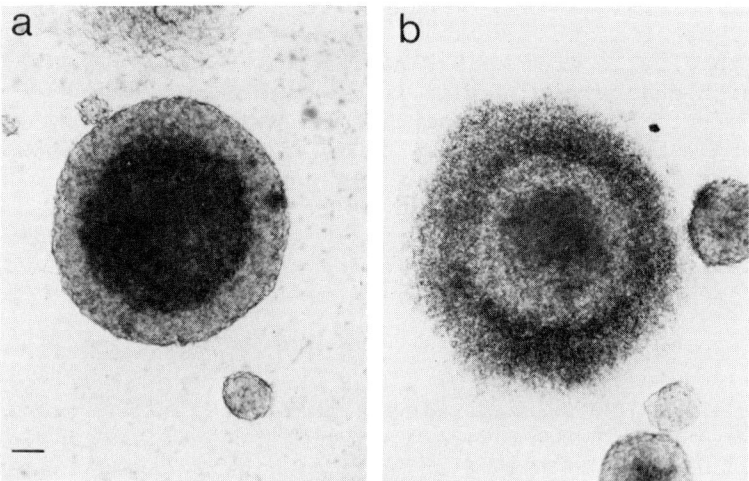

Figure 2. PCC4.azal aggregates (a) after six days on 8 um Nucleopore filter and (b) one day after attachment to the surface of a culture dish. Scale marker = 100 um.

nent nucleoli of, embryonal carcinoma cells (see Fig. 4a). These cells were seen growing as a group, usually on only one side of a clump. Subsequently, it was found that the incidence of surviving clumps giving rise to differentiated cell outgrowths could be increased to 75% by extended periods on Nucleopore filters (Table 1). On the other hand, the percentage of viable clumps was reduced by half following these prolonged treatments.

To determine whether the appearance of differentiated cells was due to the formation of aggregates or to some peculiar property of the Nucleopore filter, these experiments were repeated using bacterial petri dishes. The result was once again the formation of cell clumps which were capable of giving rise to differentiated cells when placed on a normal tissue culture surface (Table 1, Expt. 2). Although the percentage of viable aggregates remained high after extended periods in the bacterial petri dish, the percentages of viable aggregates giving rise to differentiated cells after one week in a regular tissue culture dish were very similar to those obtained by the Nucleopore filter technique.

In a second set of experiments, aggregates formed in a bacterial petri dish in DME medium were scored for outgrowth

TABLE 1

PRODUCTION OF DIFFERENTIATED CELLS BY PCC4.azal AGGREGATES

Experiment	Substratum	Time to form Aggregates (weeks)	% of Aggregates Surviving	% of Surviving Aggregates with Differentiated Cells
1	Nucleopore Filter[a]	1	97.9	37.2
		2	56.3	64.8
		4	50.0	75.0
2	Bacterial Petri Dish[b]	1	79.2	43.4
		2	94.8	74.7
		4	96.9	73.1

[a]PCC4.azal cells were placed on 8 μm Nucleopore filters in DME medium in 35 mm tissue culture dishes for the indicated periods of time. Aggregates were then individually placed in wells of a Falcon microtest dish (No. 3040) containing DME medium and scored for the presence of differentiated cells after seven days. Ninety-six aggregates were scored in each case.

[b]Conditions were as above, except that aggregates were formed in 60 mm bacterial Petri dishes.

of differentiated cells within 24 hours of transfer to a tissue culture flask containing either DME or, for reasons to be discussed below, NCTC-109 (Evans et al., 1964) medium. Table 2 illustrates that cells with a morphology different

TABLE 2

APPEARANCE OF DIFFERENTIATED CELLS IN AGGREGATE OUTGROWTHS TWENTY-FOUR HOURS AFTER TRANSFER TO TISSUE CULTURE FLASKS

Time in Bacterial Petri Dish (weeks)	Outgrowth Medium	% of Aggregate Outgrowths with Differentiated Cells[a]
1	NCTC-109	4.5
	DME	2.0
2	NCTC-109	14.5
	DME	22.5
4	NCTC-109	52.0
	DME	37.0

[a]Two hundred aggregates were scored in each case.

from embryonal carcinoma cells can indeed be detected within a day of attachment of the aggregates to the culture vessel surface. Furthermore, regardless of the medium used, the number of aggregates with differentiated cells increased proportionately with time in the bacterial petri dish (Table 2 and Fig. 3). This does not necessarily indicate that larger aggregates are more likely to give rise to differentiated cells, since the aggregates do not increase visibly in size between the second and fourth weeks in the bacterial culture dish. In fact, during scoring it was noted that many of the largest aggregates did not give rise to differentiated cells, even after seven days of culture in a tissue culture dish, by which time there was ample outgrowth of embryonal carcinoma cells. On the other hand, extrapolation of the line connecting the points in Fig. 3 suggests that aggregates only become capable of producing differentiated cells after seven days of culture (i.e., six days in a bacterial petri dish and one day in a tissue culture dish. If PCC4.azal cells grow as rapidly in aggregates as they do in monolayers (i.e., doubling time of about 10 hours), it can be estimated that the smallest aggregates capable of differentiating would contain as many as 50,000 cells. It would, therefore, appear that aggregates of PCC4.azal cells must first reach a certain critical mass before they are capable of differentiating, but after that mass is reached, other factors, including time spent in aggregate form, determine whether differentiation will take place or not.

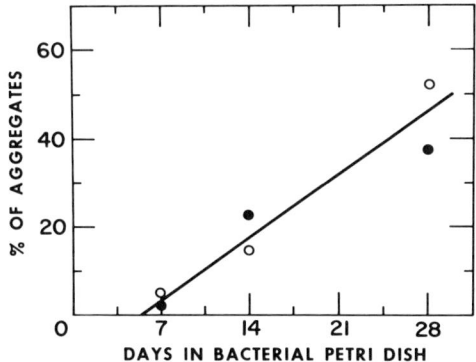

Figure 3. Correlation between duration of incubation of aggregates in a bacterial petri dish and percentage showing differentiated outgrowths. Aggregates were monitored 24 hours after transfer to a tissue culture flask containing (○) NCTC-109 or (●) DME medium.

Since differentiated cells are visible so soon after attachment of the aggregates to the culture dish, it is very likely that differentiation actually takes place within the free floating aggregates, as has been demonstrated by Rosenthal et al. (1970) and Kahan and Ephrussi (1970) with aggregates produced *in vitro* by their cell lines.

ENRICHMENT OF DIFFERENTIATED CELLS DEVELOPING FROM PCC4.azal AGGREGATES

Although most PCC4.azal aggregates give rise to differentiated cells, the vast majority of the cells in the cultures maintain the embryonal carcinoma morphology. This makes characterization of the differentiated cell types present very difficult. Therefore, efforts were made to reduce the proportion of embryonal carcinoma cells in the culture. Two procedures were found to be somewhat effective, although neither was completely satisfactory insofar as long periods of time (a month or more) are required and embryonal carcinoma cells are not completely eliminated. The first makes use of the fact that most PCC4.azal cells do not survive at high densities (Jakob et al., 1973). In this procedure, aggregates are plated in small tissue culture dishes or flasks and are maintained without transfer for a month or more. Many of the embryonal carcinoma cells lyse as high cell densities are reached. Under these conditions, non-embryonal carcinoma cell types come to predominate, and restrict the embryonal carcinoma cells to small clusters (e.g., see Fig. 4b). One disadvantage of this procedure is that the sustained period of culture at high density tends to have adverse effects upon some of the differentiated cell types as well. Another shortcoming is that some embryonal carcinoma cells survive these culture conditions, and, upon transfer, resume their very rapid growth and become predominant once again.

The second procedure involves the use of NCTC-109 medium. It has been noted that cells with an embryonal carcinoma morphology do not persist when blastocysts are cultured in this medium (Sherman, 1975c). Similarly, when PCC4.azal cells are cultured in the presence of NCTC-109 medium, they initially grow very rapidly, as they do in DME medium (doubling time of about 10 hours), but after 24 to 48 hours, the doubling time tapers off drastically to 24 hours or more. Pure cultures of PCC4.azal embryonal carcinoma cells cannot be maintained in this medium; clones appear only rarely when seeded at 500 cells/60 mm culture dish (see Table 4), and aggregates will not form when cells are maintained in bacterial petri dishes in NCTC-109 medium. When aggregates grown in DME medium are transferred to culture dishes containing NCTC-109 medium, the production of differentiated cells approximates that observed when DME medium is used (Table 2), but the results is a more rapid loss of embryonal carcinoma cells. Once again, however, some embryonal carcinoma cells persist in these mixed cultures, and grow rapidly when cultures are split.

To date, the most reliable procedure for obtaining cultures enriched in differentiated cells involves the development of PCC4.azal aggregates for two weeks in bacterial Petri dishes followed by transfer to tissue culture dishes in NCTC-109 medium. After one month in NCTC-109 medium, 75% or more of the cells remaining in the culture appear to be differentiated by morphological criteria.

PROPERTIES OF DIFFERENTIATED CELLS DERIVED FROM PCC4.azal AGGREGATES

As mentioned above, the first non-embryonal carcinoma cell type to grow out of PCC4.azal aggregates has a fibroblastic morphology (Fig. 4a). Although epithelioid cells are rarely seen initially, they usually appear after a week or more of culture. A number of epithelioid subtypes can be discerned (e.g. Fig. 4b to d). The variety illustrated in Fig. 4b often becomes the most predominant differentiated cell type. Cells with long cytoplasmic processes are also in evidence (Fig. 4c); these may well be neuronal. Small clusters of giant cells resembling trophoblast do appear (Fig. 4f), but only occasionally. Although the variety of cell types observed in any culture derived from PCC4.azal aggregates is unpredictable, it is evident that PCC4.azal embryonal carcinoma cells have the same multipotential capacity under the appropriate *in vitro* conditions as do other embryonal carcinoma cell lines.

None of the differentiated cell types observed in these cultures appear to divide quickly. In one study, for example, a culture containing mainly differentiated cells had an average doubling time in NCTC-109 medium of 32 hours. The differentiated cells in this culture had not lost resistance to high levels of 8-azaguanine (45 µg/ml) shown by the parental PCC4.azal embryonal carcinoma cells. Since cultures have not yet been totally freed of embryonal carcinoma cells, assignment of chromosome numbers to differentiated cells is only tentative. Preliminary indications, however, are that even after 2.5 months of culture, differentiated cells possess chromosome numbers in the diploid range, although with a

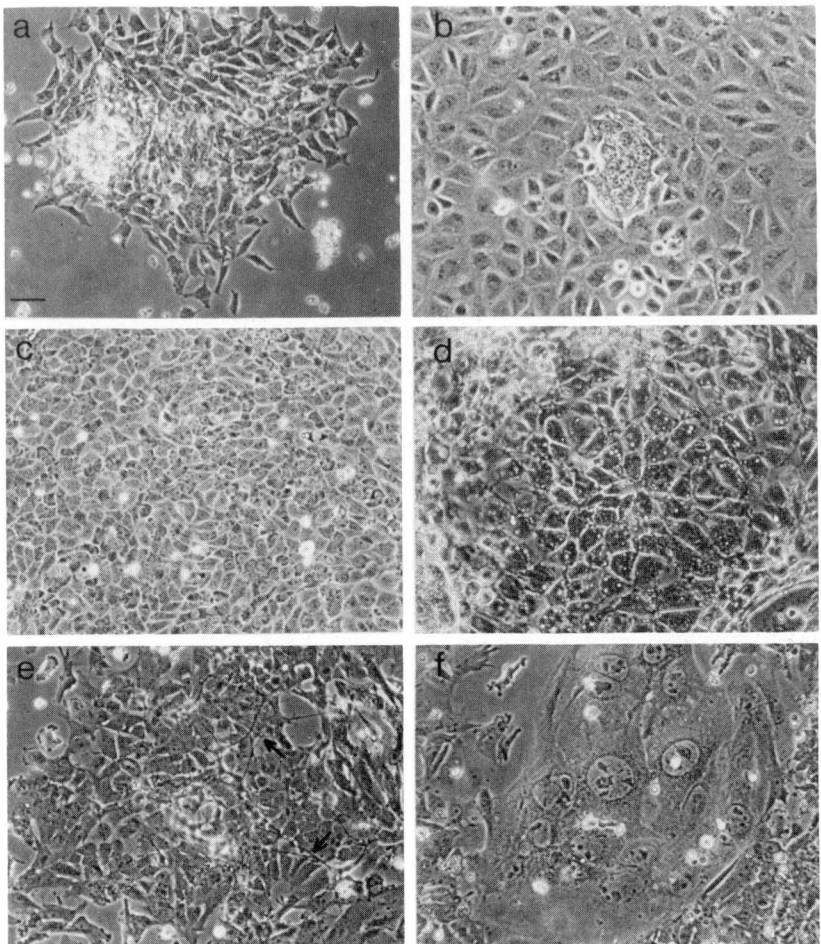

Figure 4. Some cellular morphologies in outgrowth cultures of PCC4.aza1 aggregates. Aggregates were generated by placing PCC4.aza1 embryonal carcinoma cells either on Nucleopore filters or bacterial culture dishes. Times of aggregate formation and culture following outgrowth are, respectively, as follows: (a) 7 days and 1 day; (b) 9 days and 58 days; (c) 7 days and 28 days; (d) 9 days and 94 days; (e) 7 days and 22 days, and (f) 14 days and 27 days. Note the cluster of embryonal carcinoma cells surrounded by epithelioid cells in (b) and the long cytoplasmic processes (see arrows) of some cells in (e). All photographs are at the same magnification. Scale marker (in a) = 50 um.

somewhat wider distribution than parental PCC4.azal embryonal carcinoma cells. This is also true of embryoid body cells derived from injection of PCC4.azal cells into recipient hosts (Jakob et al., 1973). Also in agreement with earlier findings on these cells and their derived embryoid bodies (Jakob et al., 1973), the majority of metaphases in PCC4.azal embryonal carcinoma cells as well as cells differentiating from PCC4.azal aggregates contain at least one metacentric chromosome.

TABLE 3

PHOSPHATASE ACTIVITIES OF PCC4.azal EMBRYONAL CARCINOMA CELL AGGREGATES AND AGGREGATE OUTGROWTH CULTURES

Sample	Specific Activity (nmoles product/min/mg protein)[a]		Alkaline Phosphatase Acid Phosphatase
	Alkaline Phosphatase	Acid Phosphatase	
PCC4.azal Embryonal Carcinoma Cells	320.2	47.3	6.8
PCC4.azal Aggregates[b]	66.6	9.2	7.3
Differentiated Outgrowths:[c]			
Culture 1	104.0	81.9	1.3
Culture 2	60.6	24.1	2.5
Culture 3	45.4	84.7	0.5
Embryoid Body Derived Culture[d]	122.0	180.5	0.7

[a]Cells were collected, homogenized and assayed for phosphatase activities using p-nitrophenyl phosphate as substrate as described previously (Sherman, 1975a).

[b]PCC4.azal aggregates used in this experiment were maintained on Nucleopore filters for six days prior to collection.

[c]Differentiated outgrowths were derived from aggregates incubated on Nucleopore filters, transferred to tissue culture dishes and maintained for 6 (Cultures 1 and 2) or 7.5 (Culture 3) weeks prior to assay. These cultures contained a variety of differentiated cell types as well as small numbers of embryonal carcinoma cells.

[d]An embryoid body derived culture obtained from Mr. R. Miller, Yale University, originally derived from the transplantable teratocarcinoma OTT 6050 (Stevens, 1970). The culture contained a variety of differentiated cell types as well as large numbers of embryonal carcinoma cells.

Initial experiments support the view that non-embryonal carcinoma cells produced by PCC4.azal aggregates differ from embryonal carcinoma cells not only morphologically, but biochemically as well. In prior histochemical studies on teratomas derived from egg cylinders, Damjanov et al. (1971) reported that in frozen sections, embryonal carcinoma cells had high levels of alkaline phosphatase activity but were devoid of acid phosphatase activity. On the other hand, adjacent differentiated cells often had less alkaline phosphatase but more acid phosphatase activity than embryonal carcinoma cells. Bernstine et al. (1973) have confirmed that embryonal carcinoma cells have high levels of alkaline phosphatase both histochemically and biochemically. Biochemical assays for alkaline and acid phosphatase activities in PCC4.azal cells and their derivatives are generally consistent with these results (Table 3). The parental PCC4.azal line has a high level of alkaline phosphatase and a high alkaline phosphatase/acid phosphatase ratio. PCC4.azal aggregates have the same high ratio, but lower specific activities of both alkaline and acid phosphatase activities, possibly due to necrosis in the center of the aggregates. Three cultures derived from PCC4.azal aggregate outgrowths six to eight weeks earlier and containing a high proportion of differentiated cells all possessed lower specific alkaline phosphatase activities than those of embryonal carcinoma cells. The alkaline/acid phosphatase ratios were in all cases markedly lower than that of the parental PCC4.azal cells and closer to that of cells growing out from cultured embryoid bodies.

Of the three differentiated outgrowth cultures in Table 3, one of them, Culture 2, possessed an electrophoretic esterase pattern that was qualitatively different from that of the parental embryonal carcinoma cells; culture 1 and 3, as well as one embryoid body culture tested, possessed qualitatively similar profiles (unpublished observations). Further biochemical characterizations are in progress.

CLONING STUDIES

PCC4.azal cells clone at a relatively high frequency, even in the absence of feeder layers or conditioned medium (Table 4). After ten days or so, the clones usually contain thousands of cells, and often closely resemble aggregates just after they have attached to the culture dish (e.g. Fig. 2b), both in shape and size. Unlike aggregates, however, very few of these clones give rise to cells which can be scored as nonembryonal carcinoma on the basis of morphology

TABLE 4

CLONING STUDIES OF PCC4.azal CELLS AND PCC4.azal
AGGREGATE OUTGROWTH CULTURES

Culture	Medium	Total Cells Plated[a]	Total Clones[b]		Clones Containing Differentiated Cells		Clones Containing *Only* Differentiated Cells	
			No.	%	No.	%	No.	%
PCC4.azal	DME	6000	1138	19.0	11	1.0	0	0
	NCTC-109	6000	7	0.1	1	14.3	0	0
Differentiated Outgrowths:[c]								
Culture 4	DME	3132	265	8.5	47	17.7	4	1.5
	NCTC-109	2610	0	0				
Culture 5	DME	5140	248	4.8	52	23.0	5	2.0
	NCTC-109	2570	0	0				
Culture 6	DME	3992	281	7.0	24	8.2	4	0.4
	NCTC-109	2495	0	0				
Culture 7	DME	6000	162	2.7	28	17.3	1	0.6
	NCTC-109	6000	10	0.2	2	20.0	1	10.0

[a] Cells were plated at a density of approximately 500 cells/60 mm culture dish.

[b] Clones containing at least 20 cells were scored after 10-14 days of culture. By this time, pure PCC4.azal clones contained many thousands of cells.

[c] Differentiated outgrowths were derived from aggregates incubated on Nucleopore filters (Cultures 4-6) or bacterial Petri dishes (Culture 7). Cloning studies were carried out after 7 (Cultures 4 and 5), 8.5 (Culture 6) and 4 weeks (Culture 7) of outgrowth. PCC4.azal and Cultures 4-6 were maintained in DME medium prior to cloning; Culture 7 was maintained in NCTC-109 medium for 3 weeks prior to cloning.

(Table 4), and even in these cases, the morphological differences are often slight. As mentioned above, cloning efficiency drops dramatically when NCTC-109 medium is used.

When cells from aggregate outgrowths are plated at very low density under the above conditions, very few clones containing *only* differentiated cells are detected (Table 4). Characteristically, these clones contain few cells and cannot be maintained. The majority of the clones contain embryonal carcinoma cells. The reduced cloning frequency (by 55-85%) of cells from aggregate outgrowths compared to the parental PCC4.azal cells no doubt reflects the presence of a high percentage of differentiated cells in the former cultures. The significant observation, however, is that between 8 and 23 percent of the clones developing from aggregate outgrowths contain not only embryonal carcinoma cells, but differentiated cells as well. The gross shape of these mixed clones is not different from that of parental PCC4.azal clones. Interestingly, in many cases, the first nonembryonal carcinoma cells observed do not resemble the earliest type to appear from aggregates (Fig. 4a); epithelioid cells are often the first in evidence. After several weeks, however, a variety of cell types can be observed.

DISCUSSION AND CONCLUSIONS

While this work was in progress, other laboratories have also been studying the potentials of embryonal carcinoma cell derived aggregates in culture, although in these cases, the embryonal carcinoma cells used were pluripotent *in vitro*. The observations of Nicolas et al.(1975) with aggregates from embryonal carcinoma cell line PCC3 are in close agreement with those reported here. These investigators generated aggregates by centrifuging embryonal cells and incubating the cell pellets at 37°. They demonstrated that the resultant aggregates must reach a critical mass before cell differentiation occurs. The data in Fig. 3 above indicates that the same applies to PCC4.azal aggregates. The first differentiated cells to migrate out from the aggregates formed by Nicolas and coworkers strongly resemble those initially observed with PCC4.azal aggregates (Fig. 4a). Finally, Nicolas et al. (1975) also found that within a month of outgrowth from embryonal carcinoma cell aggregates, differentiated cells still maintained the diploid number of chromosomes.

Our observations differ in some respects from those recently described by Martin and Evans (1975 and also this volume), who found that subclones of the pluripotent SIKR teratoma line have a tendency to form rounded aggregates spontaneously. These contain an outer layer of cells which they believe to be parietal endoderm. When the aggregates attach to a substratum, the first differentiated cell type to outgrow is epithelioid, although a variety of other cell morphologies appear later. The properties of these SIKR cell derived aggregates have led Martin and Evans to liken their pattern of development to that in early embryogenesis.

Taken together, the observations of Nicolas et al.(1975), Martin and Evans (1975) and those described above strongly point to the requirement for a three dimensional matrix before differentiation of embryonal carcinoma cells can occur. As pointed out in the introduction, initial differentiation of the ICM of the blastocyst *in vitro* also takes place in a solid clump of cells, so that a three dimensional matix may in some way trigger differentiation of the early embryo as well. It should be noted, however, that the ICM clusters giving rise to differentiated cells generally contain considerably fewer cells than the estimated critical mass of PCC4.azaI aggregates. Furthermore, although embryonal carcinoma cell derived aggregates may develop in a pattern resembling that of the ICM (Martin and Evans, 1975), studies with PCC3 (Nicolas et al., 1975) and PCC4.azaI aggregates make it clear that other pathways of differentiation are available to embryonal carcinoma cells.

The experiments described here suggest that PCC4.azaI cells show more potential for differentiation *in vivo* than they do under normal culture conditions only because they do not normally form solid aggregates *in vitro*. However, clones of embryonal carcinoma cells derived from aggregate outgrowths have the same gross appearance as those of parental PCC4.azaI embryonal carcinoma cell clones and yet the latter rarely give rise to differentiated cell types while the former often do. This would indicate that PCC4.azaI embryonal carcinoma cells need only receive an initial stimulus to differentiate by their presence in an aggregate. Thereafter, many of these cells and their progeny retain the ability to undergo the necessary steps leading to cell differentiation.

ACKNOWLEDGEMENT

These experiments were conceived as a result of discussions with Dr. Thierry Boon. I wish to thank him for suggestions and for the gift PCC4.azaI cells.

REFERENCES

Artzt, K., Dubois, P., Bennett, D., Condamine, H., Babinet,

C., Jacob, F. (1973). Surface antigens common to mouse cleavage embryos and primitive teratocarcinoma cells in culture. Proc. Nat. Acad. Sci. U. S. 70, 2998.

Bell, K. E., Sherman, M. I. (1973). Enzyme markers of mouse yolk sac differentiation. Develop. Biol. 33, 38.

Bernstine, E. G., Hooper, M. L., Grandchamp, S., Ephrussi, B. (1973). Alkaline phosphatase activity in mouse teratoma. Proc. Nat. Acad. Sci. U. S. 70, 3899.

Damjanov, I., Solter, D., Skreb, N. (1971). Enzyme histochemistry of experimental embryo-derived teratocarcinomas. Z. Krebsforsch. 76, 249.

Edidin, M., Patthey, H. L., McGuire, E. J., Sheffield, W. D. (1971). An antiserum to "embryoid body" tumor cells that reacts with normal mouse embryos. In "Embryonic and Fetal Antigens in Cancer" (Eds. N. G. Anderson and J. H. Coggin, Jr.). ORNL, Oak Ridge, p. 239.

Evans, M. J. (1972). The isolation and properties of a clonal tissue culture strain of pluripotent mouse teratoma cells. J. Embryol. Exp. Morph. 28, 163.

Evans, V. J., Bryant, J. C., Kerr, H. A., Schilling, E. L. (1964). Chemically defined media for cultivation of long term strains from four mammalian species. Exp. Cell Res. 36, 439.

Finch, B. W., Ephrussi, B. (1967). Retention of multiple developmental potentialities by cells of a mouse testicular teratocarcinoma during prolonged culture *in vitro* and their extinction upon hybridization with cells of permanent lines. Proc. Nat. Acad. Sci. U. S. 57, 615.

Hsu, Y.-C. (1971). Post-blastocyst differentiation *in vitro*. Nature 231, 100.

Hsu, Y.-C. (1973). Differentiation *in vitro* of mouse embryos to the stage of early somite. Develop. Biol. 33, 403.

Hsu, Y.-C. Baskar, J., Stevens, L. C., Rash, J. E. (1974). Development *in vitro* of mouse embryos from the two-cell stage to the early somite stage. J. Embryol. Exp. Morph. 31, 235.

Jakob, H., Boon, T., Gaillard, J., Nicolas, J.-F., Jacob, F. (1973). Teratocarcinome de la souris: isolement, culture et proprietes de cellules a potentialites multiples. Ann Microbiol. (Inst. Pasteur) 124B, 269.

Kahan, B. W., Ephrussi, B. (1970). Developmental potentialities of clonal *in vitro* culture of mouse testicular teratoma. J. Nat. Cancer Inst. 44, 1015.

Kahan, B., Levine, L. (1971). The occurance of a serum α_1 protein in developing mice and murine hepatomas and teratomas. Cancer Res. 31, 930.

Martin, G. R., Evans, M. J. (1974). The morphology and

growth of a pluripotent teratocarcinoma line and its derivatives in tissue culture. Cell 2, 163.

Martin, G. R., Evans, M. J. (1975). The differentiation of clonal lines of teratocarcinoma cells: formation of embryoid bodies *in vitro*. Proc. Nat. Acad. Sci. U. S., 72, 1441.

Miller, R. A., Ruddle, F. H., Sherman, M. I. (1975). Surface antigens of blastocyst-derived cell lines. This volume, p. 123.

Nicolas, J.-F., Dubois, P., Jakob, H., Gaillard, J., Jacob, B. (1975). Teratocarcinome de la souris: differenciation en culture d'une ligne de cellules primitives a potentialites multiples. Ann. Microbiol. (Inst. Pasteur) 126A, 3.

Rosenthal, M. D., Wishnow, R. M., Sato, G. H. (1970). *In vitro* growth and differentiation of clonal populations of multipotential mouse cells derived from a transplantable testicular teratoma. J. Nat. Cancer Inst. 44, 1001.

Sherman, M. I. (1972). Biochemistry of differentiation of mouse trophoblast: esterase. Exp. Cell Res. 75, 449.

Sherman, M. I. (1975a). Long term culture of cells derived from mouse blastocysts. Differentiation 3, 51.

Sherman, M. I. (1975b). The role of cell-cell interactions during early mouse embryogenesis. In "The Early Development of Mammals" (Eds. M. Balls and A. E. Wild). Cambridge University Press, London, p. 145.

Sherman, M. I. (1975c). The culture of cells derived from mouse blastocysts. Cell 5, 343.

Stevens, L. C. (1958). Studies on transplantable testicular teratomas of strain 129 mice. J. Nat. Cancer Inst. 20, 1257.

Stevens, L. C. (1968). The development of teratomas from intratesticular grafts of tubal mouse eggs. J. Embryol. Exp. Morph. 20, 329.

Stevens, L. C. (1970). The development of transplantable teratocarcinomas from intratesticular grafts of pre- and postimplantation mouse embryos. Develop. Biol. 37, 364.

VI.
Properties of Embryonal Carcinoma Cells

ULTRASTRUCTURE OF MURINE TERATOCARCINOMAS

IVAN DAMJANOV[1] AND DAVOR SOLTER[2]

[1]Department of Pathology
University of Connecticut School of Medicine,
Farmington, Ct.   06032 and
[2]The Wistar Institute of Anatomy and Biology
Philadelphia, Pa.   19104

Electron microscopy and histochemistry have so far proved not only to be most valuable techniques for elucidating various intriguing aspects of teratocarcinoma formation, but have also lead us to pose many new and rather provocative questions, most of which are still unanswered. In this article we will review briefly the contributions made through the fine structural analysis of teratocarcinomas and at the same time discuss the new vistas outlined by the knowledge accumulated to date.

ORIGIN OF TERATOCARCINOMAS

Spontaneous murine teratomas and teratocarcinomas originate either from cells in the testes (Stevens, 1967) or ovaries (Stevens and Varnum, 1974). Experimentally it was possible to produce the same tumors from male fetal genital ridges transplanted into adult testes (Stevens, 1964) or from presomitic embryos transplanted to extrauterine sites (Stevens, 1968, 1970; Solter et al., 1970). The fact that teratomas originate from organs that contain germ cells is in accord with experience from human pathology and with the hypothesis of the germ cell origin of teratomas. Comparative electron microscopic studies performed on fetal mouse testes and testicular teratocarcinomas (Pierce and Beals, 1964) revealed a remarkable similarity between the tumor stem cells and primordial germ cells. Finally, sequential analysis of cytological events in transplanted fetal male genital ridges confirmed the previous notion that the tumors originated from primordial germ cells (Pierce et al., 1967).

Teratocarcinomas can be experimentally produced from early mouse embryos transplanted to extrauterine sites. The malignant stem cells of the tumor are direct descendants of the transplanted cells of the presomitic embryo. Theoretically these cells could stem either from the endodermal or ectodermal germ layer. On the basis of ultrastructural analyses, we have proposed that the tumors most probably originate from the ectodermal layer (Damjanov et al., 1971a). In contrast to the highly specialized endodermal cells which perform a nutritive function in the two-layered egg-cylinder, the ectodermal cells have the appearance of undifferentiated cells (Solter et al., 1970b) and, therefore, closely resemble the undifferentiated stem cells of teratocarcinomas (Fig. 1).

Teratomas stemming from ovaries of LT mice are undoubtedly derived from parthenogenetically activated ova (Stevens and Varnum, 1974). The ovaries of LT mice contain parthenotes in all stages of development up to the egg-cylinder. From the stage of egg-cylinder, the development of parthenotes becomes disorganized and teratomas form. In analyzing the stem cells of ovarian teratomas, we have noted that these cells are also ultrastructurally similar to the ectodermal cells of the normal mouse egg-cylinder (Damjanov et al., 1975) and do not differ from the stem cells in embryo-derived, or testicular, teratocarcinomas. Stem cells of ovarian teratocarcinomas are, however, ultrastructurally dissimilar from the ovum, their cell of origin, as well as from the cells of early cleavage stage embryos which develop from the parthenogenetically activated ova preceding the formation of teratomas.

In summary, from the ultrastructural data on stem cells of testicular, ovarian and embryo-derived teratocarcinomas, it becomes evident that these cells resemble each other despite their different origins. It is tempting to assume that there is a common developmental pathway leading from germ cells or embryos to teratocarcinoma. The pluripotent stem cell equivalent to ectodermal cells from 7 day mouse egg-cylinders would, according to this hypothesis, represent the common ancestral cell of all murine teratocarcinomas. Exper-

---

ABBREVIATIONS AND TERMINOLOGY

APase - alkaline phosphatase.

The day of observation of the sperm plug is considered day zero of pregnancy.

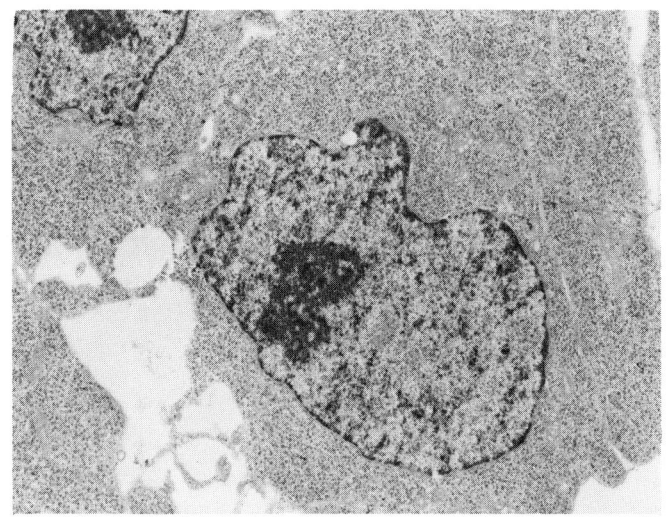

Figure 1. Ectodermal cells from a 7 day mouse egg cylinder. The nucleus contains a large nucleolus and euchromatin predominates over heterochromatin. The cytoplasm of these cells is replete with free ribosomes. There are few mitochondria. Other cytoplasmic organelles are inconspicuous. x 8500.

iments with embryo-derived teratocarcinomas form the basis and a primary confirmation of this presumption. Irrespective of the age of the presomitic embryo transplanted to extrauterine sites, the transplant will develop to the stage of egg-cylinder and then become disorganized (Stevens, 1968). If one transplants embryos that have passed the stage of egg-cylinder formation, one never obtains teratocarcinomas (tumors which by definition contain undifferentiated, pluripotent, 'malignant' stem cells) but only teratomas (Damjanov et al., 1971b). This indicates that embryonic cells that have passed the stage of egg-cylinder lose their pluripotency, become more developmentally restricted and cannot proliferate and retain their undifferentiated character. Analysis of ovarian teratomas lends further support to the hypothesis that teratocarcinomas form following disorganization of egg-cylinder stage embryos (Stevens and Varnum, 1974).

During the early stages of teratocarcinoma formation in testicular seminiferous tubules, egg-cylinder-like structures are not observed. Although this fact does not fully support the proposed unified concept of teratoma formation, it does not disprove it either, because the intratubular tumor cells more closely resemble ectodermal embryonic cells than the primordial germ cells from which they originate. It has to be noted however, that testicular teratocarcinomas can form embryoid bodies upon subsequent passage in other sites which recapitulate many of the essential developmental events noted in the normal mouse embryo (Pierce et al., 1960). Small embryoid bodies containing undifferentiated embryonic stem cells will, upon subcutaneous transplantation, give rise to teratocarcinomas. Large cystic embryoid bodies that have passed the critical stage of developmental pluripotency form only benign teratomas, like normal embryos that have passed the egg-cylinder stage of development.

NATURE OF STEM CELLS

Approximately 50% of all 7 day old mouse egg-cylinders transplanted under the kidney capsule of isogenic adult recipients will give rise to malignant tumors, i.e. teratocarcinomas. The fact that malignant tumors develop from untreated 'normal' embryos, will be dealt with in more detail in one of the subsequent papers in this volume (Solter et al., 1975). We would like to point out here that this tumor model poses some intriguing questions. If one transplants normal embryonic cells and obtains malignant tumors from them, where would one draw the line between normal and neoplastic development? Do the embryonic cells undergo spontaneous malignant transformation? In what respect do the 'malignant' stem cells differ from the "normal" embryonic cells from which they stem?

Our ultrastructural studies have so far revealed only similarities and almost no significant differences between normal ectodermal cells and their malignant counterparts, i.e. the stem cells from teratocarcinomas. With the aid of a trained cytopathologist, we have attempted to apply the standard cytologic criteria for malignant cells to distinguish between tumor cells and their normal embryonic equivalents, but to no avail (Dominis et al., 1975). Most of the embryo derived teratocarcinomas have a normal diploid karyotype (Dunn and Stevens, 1970). Upon retransplantation the number of chromosomes may remain stable and diploid or may change concurrently with the progression of the tumor into a more malignant, faster growing one. Stem cells of these aneuploid or polyploid tumor lines acquire the usual cytologic characteristics of malignancy (Dominis et al., 1975), although they still retain the capacity to differentiate into nonproliferating, 'benign' somatic tissues. Morphologic and chromosomal changes noticed on serial transplantation of teratocarcinomas indicate that events, possibly equivalent to so-called malignant transformation *in vitro*, can occur. These changes appear to be secondary phenomena and cannot be used to explain the transition of normal embryonic cells into stem cells of teratocarcinomas.

In searching for a mechanism that will cause the transition of transplanted embryonic cells into the stem cells of teratocarcinoma, one cannot bypass the oncogene theory of Todaro and Huebner (1972). There are indeed endogeneous viral particles present both in early mouse embryos (Chase and Piko, 1973; Biczysko et al., 1973) as well as in teratocarcinomas. The role of these endogeneous viral particles is not known, but it has been hypothesized that they might play a role in neoplasia (Todaro, 1974). In mouse embryos, the appearance and number of viral particles depends on embryonic age (Chase and Piko, 1973; Biczysko et al., 1973). The developmental stage of a given cell in a teratocarcinoma might also be the crucial factor determining whether one will or will not detect endogeneous viral particles. Endogenous viral particles have been detected ultrastructurally both in the undifferentiated stem cells as well as in the somatic cells derived from them (Damjanov et al., 1975).

Teratocarcinomas derived from transplanted presomitic embryos are undoubtedly malignant. The tumors have a tendency for unrestricted growth and finally cause the death of the host either due to emaciation or metastasis. Generally, the only malignant, i.e. rapidly proliferating, element in these tumors are their undifferentiated stem cells. As mentioned above we were unable to detect any significant differences between the tumor stem cells and normal embryonic ectodermal cells. Neither could we define or detect the point of transition or transformation of embryonic into tumor cells. Lacking evidence for such a transformation we are left with the assumption that the tumor stem cells are actually the undifferentiated embryonic cells themselves equivalent to pluripotent ectodermal cells from the 7 day old mouse egg-cylinders. Under proper circumstances these cells will differentiate and form nonproliferating somatic tissues. Lacking the proper stimuli, or having escaped the proper control mechanisms, they will proliferate and form malignant tumors. Malignancy appears to be just one of their possible developmental pathways.

With this presumption of the identity of ectodermal embryonic cells and tumor stem cells in mind, we have searched for biochemical markers common to both of these cells. We found that a rather nonspecific enzyme, alkaline phosphatase (APase), can be demonstrated histochemically both in the ectodermal germ layer (Solter et al., 1973) and in the nests of stem cells in teratocarcinomas (Damjanov et al., 1971c). The enzyme activity is restricted to the inner cell mass of the blastocyst, and later to the ectoderm and mesoderm; the trophoblast and the endoderm do not show any APase activity. Although the enzyme activity in various tissues can be demonstrated in different intracellular locations (Cutler et al., 1974), in the stem cells of teratocarcinoma the activity is localized exclusively on the cell membrane (Fig. 2; Damjanov, Cutler and Solter, in preparation). The same intracellular distribution of APase can be visualized in the ectodermal cells of the mouse egg-cylinder. In the process of differentiation both the ectodermal cells and the tumor stem cells lose cell membrane associated APase activity. In some of the somatic tissues formed from tumor stem cells or in the somewhat older normal embryo, APase reappears. However, Bernstine et al. (1973) have provided evidence that the activities in teratocarcinoma stem cells and other somatic tissues may be different isozymes of APase. The presence of the same enzyme, in the same cytoplasmic lo-

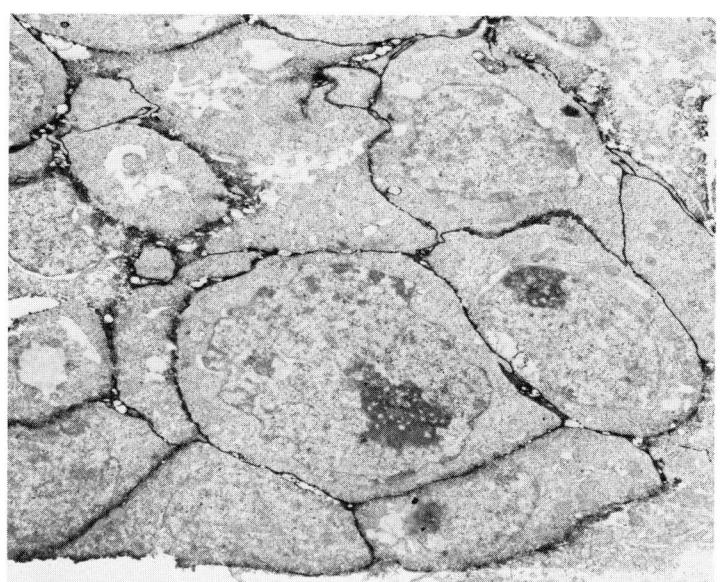

Figure 2. Alkaline phosphatase activity in stem cells of a transplantable embryo derived teratocarcinoma. Note the activity of alkaline phosphatase (dark deposits) along the cell membranes. x 6500.

cation, in ectodermal cells and the tumor stem cells has added some additional support to our hypothesis that these cells are identical.

SOMATIC TISSUES IN TERATOCARCINOMAS

In addition to rapidly proliferating stem cells, teratomas contain various somatic tissues. It was proven beyond any doubt by Kleinsmith and Pierce (1964) that the stem cells are the progenitors of all the somatic tissues in the tumors. In contrast to the rapid proliferation of stem cells, the somatic tissues are mostly quiescent or slow growing. Tumors devoid of undifferentiated stem cells never attain a substantial size, although most of these teratomas grow during the life span of the host at a rate comparable to the growth of the mice bearing the tumor. In teratocarcinomas, new somatic tissues are constantly formed from differentiating stem cells at an unpredictable rate and in an unpredictable manner. The only exceptions are the so called 'monophyletic teratocarcinomas', tumors with the capacity for differentiation restricted to one or a limited number of somatic tissues (Stevens, 1958). The fact that stem cells of monophyletic teratocarcinomas give rise to only one form, or to a restricted number of somatic tissues, indicates that they have differentiated beyond the pluripotent stage.

The only regularly malignant derivative of pluripotent stem cells of teratocarcinoma is the so-called yolk sac carcinoma (Pierce et al., 1962). It is of interest to note that malignant yolk sac carcinoma develops only from teratocarcinoma stem cells or embryonic ectodermal cells (Damjanov and Solter, 1973), i.e. cells considered to be malignant or cells capable of giving rise to malignant tumor cells. Yolk sac carcinoma cannot be grown from the parietal yolk sac or from the extraembryonic portions of the egg-cylinder (Solter and Damjanov, 1973), i.e. cells that have a programmed finite life-span.

The somatic tissues formed in teratomas show remarkable variation with regard to structural organization, maturity and homogeneity. Tissues formed through complex morphogenetic interactions (e.g. lung or kidney) are not found within teratocarcinomas. Simple structures and even functional units are, however, present, but it is not always possible to assess their functional integrity. Simple morphologic events, such as enhondral or intramembranous ossification are characterized by the usual histological and histochemical features (Damjanov et al., 1971c). Synapes between nerve cells are regularly formed and do not differ from similar structures in the normal brain. Glio-vascular junctions are also present in the neural portions of the tumors. In most cases, even the blood vessels inside such neural portions (e.g. Fig. 3) have the typical non-fenestrated endothelium, have tight junctions between endothelial cells and contain few micropinocytotic vesicles (Reese and Karnovsky, 1967), as one would expect in normal brain.

The functional maturity of somatic tissues in teratomas is hard to assess because the histologic polymorphism precludes chemical analysis. We have therefore approached this problem with a monophyletic teratocarcinoma with the capacity for differentiation restricted to neuroectodermal tissues. The tumor contained various neural and neuroglial elements at various levels of maturity in addition to its stem cells (Damjanov et al., 1973). Despite a remarkable cytological maturity in many cells, there was very little histological organization or even pattern formation. Nevertheless the tumor contained on polyacrylamide gel electrophoresis all of the rapidly migrating anodic bands typical of adult mouse brain, in general displaying a remarkable similarity with the

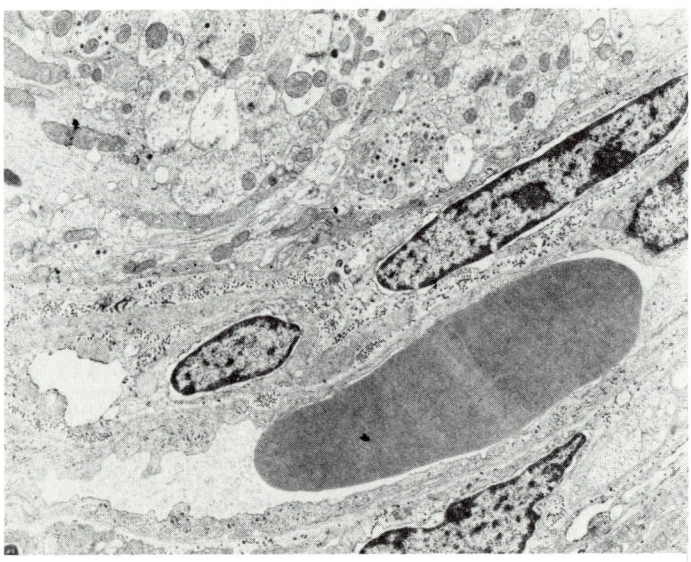

Figure 3. Blood vessel in a neural portion of a teratocarcinoma. This vessel is characteristic of those seen in adult mouse brain, indicating that normal histo-vascular relations are formed. x 7000.

protein pattern of the brain. The tumor also contained relatively large quantities of 5-hydroxytryptamine and 5-hydroxyindol acetic acid, two biogenic amines known for their specific role in neurotransmission (Jakupcevic et al., 1974). Although biogenic amines occur in other tissues as well, their presence in the neural tissue coupled with the presence of monoaminooxidase activity indicates that one is dealing with functionally active tissue and that some neurotransmission most probably occurs within the tumor. We were, however, unable to demonstrate the existence of an active transport of biogenic amines from the tumor into the blood. The absence of this mechanism could be somehow related to the histological immaturity of the neural tissue formed in the tumor.

Aberrant or defective differentiation is even harder to assess as there are no criteria for making such a judgement. We have, for example, noted that embryoid bodies formed *in vitro* from stem cells of a testicular carcinoma contain in their outer layer intermixed cells of both parietal and visceral endoderm (unpublished observations). This phenomenon is, however, observed to occur in normal mouse blastocysts cultured *in vitro* to the egg-cylinder stage (Solter et al., 1974) and is most probably an effect of *in vitro* conditions. The presence of fenestrated blood vessels in the neural segments of teratocarcinoma could be a sign of a defective structure, but, on the other hand, there are even portions of the normal adult brain that have such vessels. The presence of phagolysosomes in various cells that normally would not contain these structures is most probably a sign of suboptimal conditions for growth and functioning rather than a sign of aberrant differentiation. On the other hand we have noticed some other phenomena which are obviously abnormal, but we are unable to assess their significance. For example, we were unable to demonstrate any APase activity, by light or ultrastructural histochemistry, in the endothelial cells of vessels within teratocarcinomas. It was noticed by O'Connor (1969) that APase cannot be visualized in many malignant brain tumors, despite the fact that it is a ubiquitious enzyme and almost invariably present in the vascular endothelium. Do teratocarcinomas affect the vessels in the same way as malignant brain tumors, or is the absence of APase activity in the blood vessels only an insignificant observation that does not merit further attention? Or should one, lacking any clear cut answers, try to pursue the little irregularities in order to find some consistency and regularity in this tumor system which is characterized by inconsistency, unpredictability and irregularity?

ACKNOWLEDGEMENTS

The authors' original work was supported by grants CA 04534, CA 10815 and Ca 17546-01 from the National Cancer Institute. One of us (D.S.) was in part supported by a fellowship from the Damon Runyon Memorial Cancer Fund (DRF-810).

REFERENCES

Bernstine, E. G., Hooper, M. L., Grandchamp, S., Ephrussi, B. (1973). Alkaline phosphatase activity in mouse teratoma. Proc. Nat. Acad. Sci. U. S. 70, 3899.
Biczysko, W., Pienkowski, M., Solter, D., Koprowski, H. (1973). Virus particles in early mouse embryos. J. Nat. Cancer Inst. 51, 1041.
Chase, D. G., Piko, L. (1973). Expression of A- and C-type particles in early mouse embryos. J. Nat. Cancer Inst. 51, 1971.
Cutler, L. S., Chandhry, A. P., Montes, M. (1974). Alkaline phosphatase activity associated with the nuclear pore in normal and neoplastic salivary gland tissue. J. Histochem. Cytochem. 22, 1113.
Damjanov, I., Katic, V., Stevens, L. C. (1975). Ultrastructure of ovarian teratomas in LT mice. Z. Krebsforsch., in press.
Damjanov, I., Solter, D. (1973). Yolk sac carcinoma grown from explanted mouse egg-cylinder. Arch. Pathol. 95, 182.
Damjanov, I., Solter, D. Belicza, M., Skreb, N. (1971a). Teratomas obtained through extrauterine growth of seven-day-old mouse embryos. J. Nat. Cancer Inst. 46, 471.
Damjanov, I., Solter, D., Serman, D. (1973). Teratocarcinoma with the capacity for differentiation restricted to neuro-ectodermal tissue. Virchow Arch. Abt. B 13, 179.
Damjanov, I., Solter, D., Skreb, N. (1971b). Teratocarcinogenesis as related to the age of embryos grafted under the kidney capsule. Wilhelm Roux' Arch. Entwickl.-Mech. Org. 167, 288.
Damjanov, I., Solter, D., Skreb, N. (1971c). Enzyme histochemistry of experimental embryo-derived teratocarcinomas. Z. Krebsforsch. 76, 249.
Dominis, M., Damjanov, I., Solter, D. (1975). Cytology of experimental teratomas and teratocarcinomas. Experientia, 31, 107.
Dunn, G. R., Stevens, L. C. (1970). Determination of sex of teratomas derived from early mouse embryos. J. Nat.

Cancer Inst. 44, 99.

Jakupcevic, M., Lackovic, Z. Damjanov, I., Butlat, M. (1974). Biogenic amines in a retransplantable neurogenic teratocarcinoma. Experientia, 30, 652.

Kleinsmith, L. J., Pierce, G. B., Jr. (1969). Multipotentiality of single embryonal carcinoma cells. Cancer Res. 24, 1544.

O'Connor, J. S., (1969). Changes in histochemical staining of brain tumor blood vessels associated with increasing malignancy. Acta Neuropath. 14, 161.

Pierce, G. B., Jr., Beals, T. F. (1964). The ultrastructure of primordial germinal cells and fetal testes and of embryonal carcinoma cells in mice. Cancer Res. 24. 1553.

Pierce, G. B., Jr., Dixon, F. J., Verney, E. L. (1960). Teratocarcinogenic and tissue forming potentialities of the cell types comprising neoplastic embryoid boides. Lab. Invest. 9, 583.

Pierce, G. B., Jr., Midgley, A. R. Jr., Ram. J. S., Feldman, J. D. (1962). Parietal yolk sac carcinoma: clue to the histogenesis of Reichert's membrane of the mouse embryo. Amer. J. Path. 41, 549.

Pierce G. B., Jr., Stevens, L. C., Nakane, P. J. (1967). Ultrastructural analysis of the early development of teratocarcinoma. J. Nat. Cancer Inst. 39, 755.

Reese, T. S., Karnovsky, M. J. (1967). The structural localization of a blood-brain barrier to exogeneous peroxidase. J. Cell Biol. 34, 207.

Solter, D., Adams, N., Damjanov, I., Koprowski, H. (1975). Control of teratocarcinogenesis. This volume, p 139.

Solter, D., Biczysko, W., Pienkowski, M., Koprowski, H. (1974). Ultrastructure of mouse egg-cylinders developed *in vitro*. Anat. Rec. 180, 263.

Solter, D., Damjanov, I. (1973). Explantation of extraembryonic parts of 7-day-old mouse egg-cylinders. Experientia 29, 701.

Solter, D., Damjanov, I., Skreb, N. (1970b). Ultrastructure of mouse egg-cylinder. Z. Anat. Entwickl.-Gesch. 132, 291.

Solter, D., Damjanov, I., Skreb, N. (1973). Distribution of hydrolytic enzymes in early rat and mouse embryos-- a reappraisal. Z. Anat. Entwickl.-Gesch. 139, 119.

Solter, D., Skreb, N., Damjanov, I. (1970a). Extrauterine growth of mouse egg-cylinders results in malignant teratoma. Nature (Lond.) 227, 503.

Stevens, L. C. (1958). Studies on transplantable testicular teratomas of strain 129 mice. J. Nat. Cancer Inst. 20,

1257.

Stevens, L. C. (1967). The biology of teratomas. Advan. Morph. 6, 1.

Stevens, L. C. (1968). The development of teratomas from intratesticular grafts of tubal mouse eggs. J. Embryol. Exp. Morph. 20, 329.

Stevens, L. C. (1970). The development of transplantable teratocarcinomas from intratesticular grafts of pre- and post-implantation mouse embryos. Develop. Biol. 21, 364.

Stevens, L. C., Varnum, D. S. (1974). The development of teratomas from parthenogenetically activated ovarian mouse eggs. Develop. Biol. 37, 369.

Todaro, G. J. (1974). Endogeneous viruses in normal and transformed cells. Symp. Soc.Develop. Biol. 32, 145.

Todaro, G. J., Huebner, R. J. (1972). The viral oncogene hypothesis: new evidence. Proc. Nat. Acad. Sci. U. S. 69, 1009.

CHROMATIN PROTEINS OF MOUSE PRIMITIVE TERATOCARCINOMA CELLS: ELECTROPHORETIC COMPARISON WITH SOMATIC TISSUES OF THE MOUSE

Jacques Jami and Jacques E. Loeb

Institut de Recherches Scientifiques sur le Cancer,
CNRS, 94800 Villejuif, France

It is generally assumed that every cell of a given organism possesses the same genome and that only certain combinations of genes are transcribed in each of the cell types. Physiological support for this view was obtained through transplantation of nuclei from adult amphibian somatic tissue into enucleated amphibian eggs which resulted in normal adult frogs (Gurdon and Uehlinger, 1966). In the past years, it was inferred from *in vitro* transcription and hybridization competition experiments that isolated chromatin (a complex which associates DNA with small amounts of RNA and a very large population of proteins) maintains the specific limited patterns of the tissues of origin (Paul and Gilmour, 1968). Therefore, isolated chromatin should contain DNA-bound molecules responsible for specific modulation of the genetic program, particularly in the various differentiated cell types.

Histones are the best defined chromatin proteins. They have large amounts of basic amino acids. The complete amino acid sequences of the various histones are either known or under study. Only five types of histones exist, and all are found in virtually every tissue, whatever the organism. *In vitro* transcription experiments strongly suggest that these histones reduce the overall DNA-template capacity (Huang and Bonner, 1962; Allfrey et al., 1963). Chromatin reconstitution experiments associating DNA both with histones and with non histone proteins from various sources suggest that the inhibitory action of histones is modulated by other molecules (Paul and Gilmour, 1968; Gilmour and Paul, 1969; 1970; Kamiyama and Wang, 1971).

Non histone chromosomal (NHC) proteins are numerous and heterogeneous. However, the electrophoretic patterns of NHC proteins are different in different tissues of the same organism (Loeb and Creuzet, 1969, 1970; Dastugue et al. 1970; Elgin and Bonner, 1970; Wang, 1971; Wu et al. 1973; MacGillivray and Rickwood, 1974). In addition, Chytil and Spelsberg (1972) have demonstrated immunochemical specificity of dehistonized chromatin from several tissues in the chick. Nonetheless, NHC protein electrophoretic profiles of different tissues are more similar than the cytoplasmic protein profiles (Loeb and Creuzet, 1970). This is not surprising, since NHC proteins must include structural and enzymatic components necessary for replication and maintenance of chromosomes and for gene transcription; one expects that most of these macromolecules are identical in all the cells of the organism. However, one might expect that some of the proteins that vary between chromatins interfere with the specific programmes required for the various types of differentiated phenotypes.

Developmental processes require a succession of activity and inactivity of groups of genes. Marushige and Osaki (1967) have examined the ability of isolated chromatin of the sea urchin embryo to support RNA synthesis by exogenous RNA polymerase. Their results suggest that more genes are available for transcription in pluteus chromatin than in blastula chromatin. While there is little difference in histone content, pluteus chromatin contains twice as many NHC proteins as does blastula chromatin.

Likewise, it might be hoped that if early mouse embryo cells contain some characteristic NHC proteins that are absent in further stages of development and in adult animals, it should be possible to detect their presence by electrophoretic separation and/or by immunologic analysis. Unfortunately, too little material is available to permit direct analysis of embryonal cells by current methods. Embryonal carcinoma cells, however, provide an alternative material for analysis of chromatin proteins of mouse embryo. The following arguments support the idea that teratocarcinoma cells are the tumoral counterpart of the cells of the early stages of embry-

---

ABBREVIATIONS AND TERMINOLOGY

RSB - reticulocyte standard buffer: 0.01 M NaCl, 1.5 mM $MgCl_2$, 0.01 M Tris-HCl pH 7.
TKM - 50 mM Tris-HCl pH 7.5, 25 mM KCl, 5 mM $MgCl_2$.
SDS - sodium dodecylsulfate.
NHC proteins - non histone chromosomal proteins.

onic development:

1) experimental teratomas can be obtained by grafting early mouse embryos into ectopic sites, particularly the OTT series of transplantable tumors (Stevens, 1970);

2) these tumors have multiple developmental potentialities, including embryoid body formation, and it is well established that pluripotentiality is retained by single cells *in vivo* (Kleinsmith and Pierce, 1964) and in tissue culture (Finch and Ephrussi, 1967; Kahan and Ephrussi, 1970; Rosenthal et al., 1970), and therefore large amounts of pluripotent cells are available;

3) surface antigen (F9 antigen) common to embryonal carcinoma cells and early embryonic cells has been demonstrated (Artzt et al., 1973). Experimental data strongly suggest that F9 antigen is the wild type allele expression of the recessive t^{12} gene present in mouse embryos unable to develop beyond the morula stage in the homozygous condition (Artzt et al., 1974). H2 antigens are not expressed by embryonal carcinoma cells (Artzt and Jacob, 1974) but appear during differentiation processes *in vitro*, while F9 antigen decreases (Nicolas et al., 1975).

Thus, analysis of chromatin proteins of embryonal carcinoma cells provides one way of approaching the identification of those proteins specific for the early mouse embryo.

As a first step, we have compared histone and non histone proteins of the chromatin of a clonal culture derived from the OTT 6050 teratocarcinoma with those found in 3 adult tissues, liver, brain and thymus from 129/Sv isogenic mice.

The teratocarcinoma cells were derived from a single cell isolated with a tapered pipette under the microscope from a culture established from a solid tumor of OTT 6050. Mice inoculated with such a clonal culture develop tumors in about one in ten cases. As already described elsewhere, the tumors grow slowly in 6-10 weeks and are well differentiated. They contain neural tissues, pigmented areas, skin, muscle, cartilage, bone, marrow, and cuboidal, glandular and ciliated epithelia. (Jami and Ritz, 1974).

Cells were grown in Dulbecco modified Eagle's medium supplemented with 10% calf serum in 100 mm Falcon dishes incubated at 37°C in a 7% CO_2 atmosphere. The cells were subcultured every two days. The cells are usually round with huge nuclei; they grow in clumps that detach from the dishes, and never become confluent. No morphological sign of cell differentiation is seen in these cultures. The cells are hypotetraploid.

Cells were disrupted with a Potter homogenizer. Efficiency of cell breakage was 95% or better. Nuclei were purified by centrifugation through a cushion of 2.3 M sucrose in TKM according to Blobel and Potter (1966). Purity of nuclei was tested by phase contrast microscopy (Fig. 1), and also by light microscopy after staining with 1% toluidine blue solu-

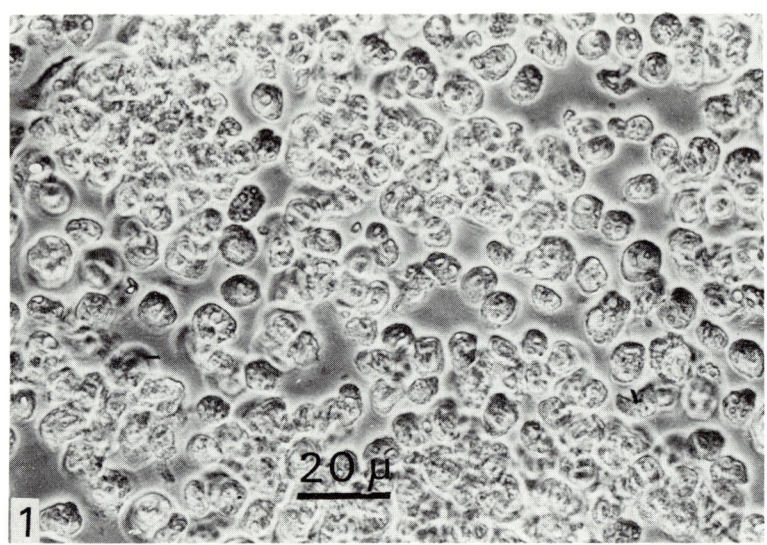

Figure 1. Purified teratoma nuclei in 0.25 M sucrose TKM by phase contrast microscopy.

tion. In the initial experiments, additional controls were carried out. Electron microscopic examination showed that the amount of contamination by membranes and cytoplasmic particles was very slight (Fig. 2). Lactic dehydrogenase and aldolase activities were absent in the nuclear pellets, indicating further that an insignificant number of unbroken cells were present in the preparations.

Nuclei from liver, thymus, and brain of 5 week-old 129/Sv strain mice were prepared in a similar way, except that cells were disrupted in 0.25 M sucrose in TKM.

All succeeding steps were identical for all types of nuclei.

Nuclear soluble proteins were extracted twice with a solution containing 0.05 M NaCl, 0.05 M NaF, 0.05 M Tris·HCl pH 7. Chromatin components were then extracted with 2 M NaCl (Wang, 1967; Loeb and Creuzet, 1969). Histones were extracted from the chromatin by adding H_2SO_4 to a concentration of 0.4 N.

DNA, total protein and histone content of isolated

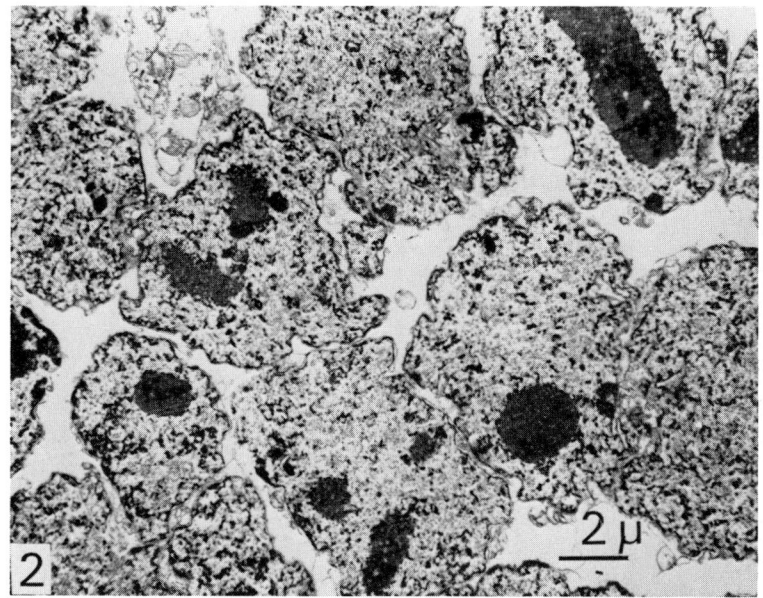

Figure 2. Thin section through a pellet of purified teratoma nuclei.

chromatin were estimated by ultraviolet absorption at 260 nm and 230 nm. The ratio of protein to DNA was higher in teratoma chromatin than in all three other types (Table 1). Since in all cases the ratio of histones to DNA is about 1.2, this shows that teratoma chromatin has more NHC proteins. This should, however, be confirmed using chromatin preparations made by different procedures.

The electrophoretic pattern obtained at pH 2.9 in 6.25 M urea (Fig. 3) was very similar for the histones from the chromatins of the teratoma and of the other tissues, except

TABLE 1

PROTEIN TO DNA RATIO IN ISOLATED CHROMATIN

	Teratoma	Liver	Brain	Thymus	L Cells
Experiment 1	3.93	2.79	2.31	1.90	2.02
Experiment 2	3.84	2.38			
Experiment 3	3.47	3.05			

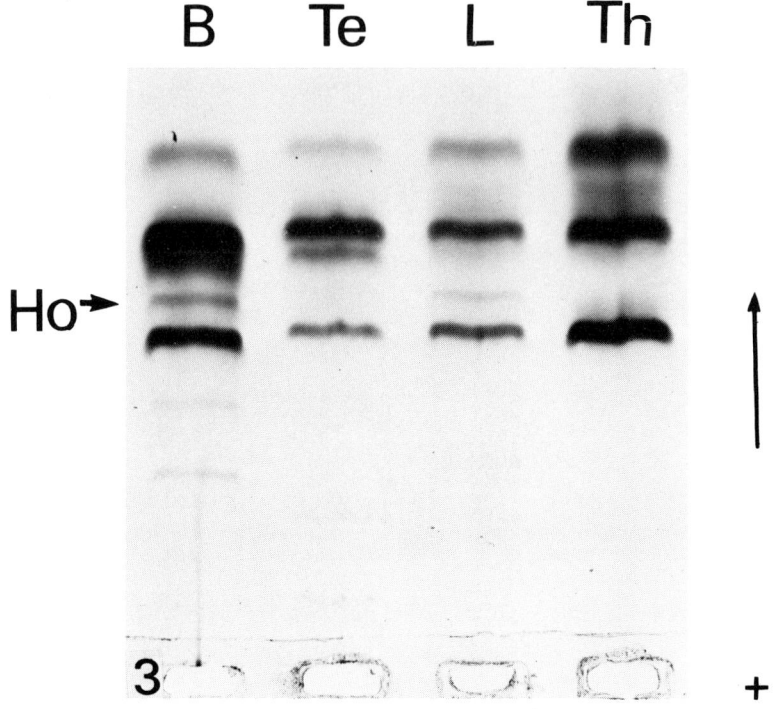

Figure 3. Electrophoresis of histones in 6.25 urea pH 2.9. Histones were extracted from isolated chromatin by adding H_2SO_4 to a concentration of 0.4 N. After centrifugation (10 000 x g, 30 min), an aliquot of the supernatant (histone fraction) was dialyzed overnight against 6.25 M urea at pH 2.9. Electrophoresis was performed in 15% polyacrylamide gels (Panyim and Chalkley 1969a). B : brain, Te : teratoma, L : liver, Th : thymus. H_O histone fraction was present in brain and liver, very weak in teratoma and absent in thymus.

for one minor band migrating after the H_1 lysine-rich histone fraction. This minor band, present in liver and brain, was very weak in teratoma. It was not detected in thymus. This band corresponds to the H_O lysine-rich histone fraction described by Panyim and Chalkley (1969b), which they found to be missing in thymus and in actively dividing cell populations, as are cells in culture. H_O was not detected when teratoma histones were analyzed in SDS gels under the same conditions as NHC proteins. Some faint bands of higher molecular weight were found in teratoma profiles. They probably are non-histone proteins entirely or partially dissolved in H₂SO₄ at the histone extraction step. A further purification or the preparation of histones by milder methods is necessary to specify which category these proteins belong to.

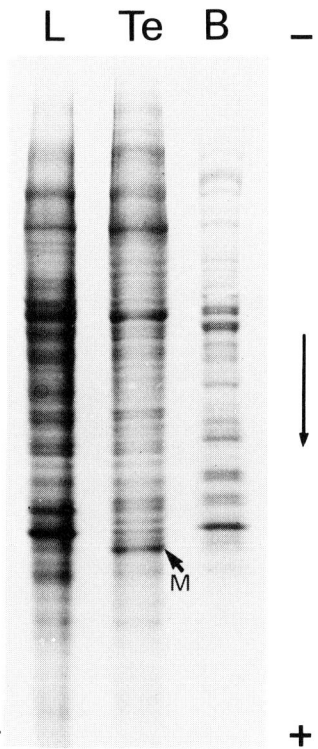

Figure 4. SDS-polyacrylamide gel electrophoresis of NHC proteins. The pellet (NHC proteins + DNA) obtained upon centrifugation of the isolated chromatin in 0.4 N H_2SO_4 was resuspended in electrophoresis sample buffer (0.125 M Tris-HCl, pH 6.8, 2% SDS, 10% glycerol and 1% 2-mercaptoethanol), dissociated by incubating 5 min in a boiling water bath with frequent stirring and dialyzed overnight against sample buffer. Electrophoresis in SDS-polyacrylamide gels was carried out according to Laemmli (1970), using slab gels in the apparatus described by Reid and Bieleski (1968). The separating gels were 10% acrylamide, and the stacking gels, 6%. Gels were stained with 0.25% Coomassie Brilliant Blue. L : liver, Te : teratoma, B : brain.

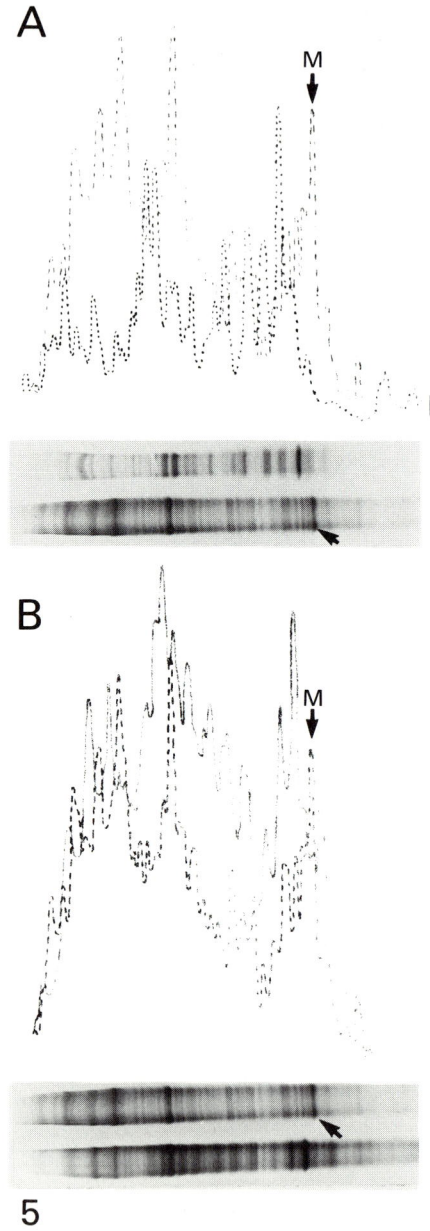

Figure 5. Densitometer scans of stained SDS-polyacrylamide gels. A. Comparison of teratoma (---) and brain (·····) profiles. B. Comparison of teratoma (---) and liver (——) profiles. The origin is on the left.

SDS gel electrophoretic diagrams of NHC proteins from teratoma and from the three tissues contained 40 to 50 bands, depending upon the resolution of the gels. At first sight, the diagrams appear broadly similar. However, systematic examination reveals clear and reproducible differences between teratoma, liver, thymus, and brain (Figs. 4 and 5).

To facilitate comparison, we have attributed a letter to 15 main bands of teratoma. Figure 6 is a schematic reproduction of the diagrams. Most of the teratoma bands were also found in one or more of the three other tissues examined. However, the M band was only conspicuous in the teratoma profile. M was indeed present in the other three tissues, but at levels about an order of magnitude lower. I band was completely absent in liver and thymus, but a very faint

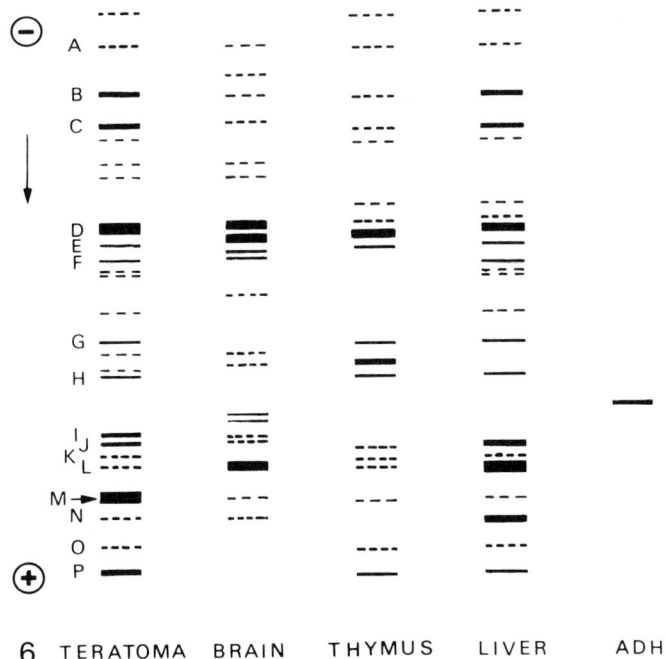

Figure 6. Simplified representation of SDS-polyacrylamide gel electrophoretic patterns of chromatin from teratoma cells, brain, thymus, and liver. Only 26 bands in teratoma were drawn, and a letter (A to O) attributed to the 15 main bands. ADH : yeast alcohol dehydrogenase, molecular weight 37,000 daltons.

doublet existed in brain at the same level. Two bands between C and D, and one just after G were only found in teratoma and brain. A doublet and a single band after F were found only in teratoma and liver.

The heterogeneity of NHC proteins in the three tissues was analyzed mainly by reference to the markers indicated for teratoma NHC proteins. In liver, L was very strong. In thymus, F and N were not visible. The diagram for brain differed most from the others, as already indicated by Wu et al. (1973). A band between A and B, and a band between F and G were only seen in this tissue; C, D, E, and F were shifted up; the band after C in other chromatins, and G, H, J, K, O, and P were absent.

Several other bands were present in the gels in addition to those indicated in Figure 6, most of which are probably identical in the four types of chromatin.

In general, the SDS polyacrylamide gel electrophoresis diagram of NHC proteins of teratoma cells differs more from that for brain than from those for liver and thymus. The group of bands I to N, beginning at the region of yeast alcohol dehydrogenase migration, and therefore having molecular weights lower than 37,000 daltons, makes possible the identification of teratoma chromatin from those of the three tissues examined.

The most striking characteristic of NHC proteins of the teratoma clone studied, however, was the conspicuous M band.

Improved methods are necessary for better characterization of these proteins, especially in the case of M, which was weakly present in other tissues. Since SDS polyacrylamide gel electrophoresis separates the proteins on the basis of individual polypeptide size, but not charge, a single band like M may represent more than one single type of protein.

It is also necessary to check that the M band is not peculiar to any rapidly growing cell culture. In two separate experiments, M was not present as a major component in NHC proteins of mouse L cells (unpublished observations). It is tempting to believe that this protein may be characteristic for teratoma cells, but this requires that its presence in independent teratoma clonal cultures be confirmed, and that its fate in differentiating teratoma cells be specified. It would also be of interest to correlate the pattern of NHC proteins of teratoma x fibroblast and teratoma x neuroblastoma cell hybrids with their phenotypes which are respectively fibroblastic (Jami et al., 1973) and neural (Jami and Diacumakos, unpublished).

ACKNOWLEDGEMENT

We thank Dr. Fanny Schapira for carrying out aldolase and lactic dehydrogenase assays.

REFERENCES

Allfrey, V. G., Littau, V. C., Mirsky, A. E. (1963). On the role of histones in regulating ribonucleic acid synthesis in the cell nucleus. Proc. Nat. Acad. Sci. U.S. 49, 414.
Artzt, K., Dubois, P., Bennett, D., Condamine, H., Babinet, C., Jacob, F. (1973). Surface antigens common to mouse cleavage embryos and primitive teratocarcinoma cells in culture. Proc. Nat. Acad. U.S. 70, 2988.
Artzt, K., Bennett, D., Jacob, F. (1974). Primitive teratocarcinoma cells express a differentiation antigen specified by a gene at the T-locus in the mouse. Proc. Nat. Acad. Sci. U.S. 71, 811.
Artzt, K., Jacob, F. (1974). Absence of serologically detectable H-2 on primitive teratocarcinoma cells in culture. Transplantation 17, 632.
Blobel, G., Potter, V. R. (1966). Nuclei from rat liver: isolation method that combines purity with high yield. Science 154, 1662.
Chytil, F., Spelsberg, T. C. (1971). Tissue differences in antigenic properties of non-histone protein - DNA complexes. Nature New Biol. 233, 215.
Dastugue, B., Tichonicky, L., Penit-Soria, J., Kruh, J. (1970). Comparaison des proteines de la chromatine, du noyau et du cytoplasme et des RNA de trois types de cellules differenciees. Bull. Soc. Chim. Biol. 52, 391.
Elgin, S. C. R., Bonner, J. (1970). Limited heterogeneity of the major nonhistone chromosomal proteins. Biochemistry 9, 4440.
Finch, B. W., Ephrussi, B. (1967). Retention of multiple developmental potentialities by cells of a mouse testicular teratocarcinoma during prolonged culture *in vitro* and their extinction upon hybridization with cells of permanent lines. Proc. Nat. Acad. Sci. U. S. 57, 615.
Gilmour, R. S., Paul, J. (1969). RNA transcribed from reconstituted nucleoprotein is similar to natural RNA. J. Mol. Biol. 40, 137.
Gilmour, R. S., Paul, J. (1970). Role of non-histone components in determining organ specificity of rabbit chromatins. FEBS Letters 9, 242.

Gurdon, J. B., Uehlinger, V. (1966). "Fertile" intestine nuclei. Nature (London) 210, 1240.

Huang, R. C. C., Bonner, J. (1962). Histone, a suppressor of chromosomal RNA synthesis. Proc. Nat. Acad. Sci. U. S. 48, 1216.

Jami, J., Failly, C., Ritz, E. (1973). Lack of expression of differentiation in mouse teratoma-fibroblast somatic cell hybrids. Exp. Cell Res. 76, 191.

Jami, J., Ritz, E. (1974). Multipotentiality of single cells of transplantable teratocarcinoma derived from mouse embryo graft. J. Nat. Cancer Inst. 52, 1547.

Kahan, B. W., Ephrussi, B. (1970). Developmental potentialities of clonal *in vitro* cultures of mouse testicular teratoma. J. Nat. Cancer Inst. 44, 1015.

Kamiyama, M. Wang, T. Y. (1971). Activated transcription from liver chromatin by nonhistone proteins. Biochim. Biophys. Acta 228, 563.

Kleinsmith, L. J., Pierce, Jr., G. B. (1964). Multipotentiality of single embryonal carcinoma cells. Cancer Res. 24, 1544.

Laemmli, U. K. (1970). Cleavage of structural proteins during the assembly of head of bacteriophage T4. Nature (London) 227, 680.

Loeb, J. E., Creuzet, C. (1969). Electrophoretic comparison of acidic proteins of chromatin from different animal tissues. FEBS Letters 5, 37.

Loeb, J. E., Creuzet, C. (1970). Comparison des proprietes electrophoretiques des proteines nucleaires de differents tissus. Bull. Soc. Chim. Biol. 52, 1007.

MacGillivray, A. J., Rickwood, D. (1974). The heterogeneity of mouse-chromatin nonhistone proteins as evidenced by two-dimensional polyacrylamide gel electrophoresis and ion-exchange chromatography. Eur. J. Biochem. 41, 181.

Marushige, K., Ozaki, H. (1967). Properties of isolated chromatin from Sea Urchin Embryo. Develop. Biol. 16, 474.

Nicolas, J.-F., Dubois, P., Jakob, H., Gaillard, J., Jacob, F. (1975). Teratocarcinome de la souris: differenciation en culture d'une lignee de cellules primitives a potentialites multiples. Ann. Microbiol. (Inst. Pasteur) 126A, 3.

Panyim, S., Chalkley, R. (1969a). High resolution acrylamide gel electrophoresis of histones. Arch. Biochem. Biophys. 130, 337.

Panyim, S., Chalkley, R. (1969b). A new histone found only in mammalian tissues with little cell division. Biochem. Biophys. Res. Comm. 37, 1042.

Paul, J., Gilmour, R. S. (1968). Organ-specific restriction of transcription in mammalian chromatin. J. Mol. Biol. 34, 305.

Reid, M. S., Bieleski, R. L. (1968). A simple apparatus for vertical flat-sheet polyacrylamide gel electrophoresis. Anal. Biochem. 22, 374.

Rosenthal, M. D., Wishnow, R. M., Sato, G. H. (1970). *In vitro* growth and differentiation of clonal populations of multipotential mouse cells derived from a transplantable testicular teratocarcinoma. J. Nat. Cancer Inst. 44, 1001.

Stevens, L. C. (1970). The development of transplantable teratocarcinomas from intratesticular graft of pre- and postimplantation mouse embryos. Develop. Biol. 21, 364.

Wang, T. Y. (1967). The isolation, properties, and possible functions of chromatin acidic proteins. J. Biol. Chem. 242, 1220.

Wang, T. Y. (1971). Tissue-specificity of nonhistone chromosomal proteins. Exp. Cell Res. 69, 217.

Wu, F. C., Elgin, C. R., Hood, L. E. (1973). Nonhistone chromosomal proteins of rat tissues. A comparative study by gel electrophoresis. Biochemistry 12, 2792.

VII.
Properties of Teratomas In Vitro

THE DIFFERENTIATION OF CLONAL TERATOCARCINOMA CELL CULTURES *IN VITRO*

MARTIN J. EVANS AND GAIL R. MARTIN[1]

Department of Anatomy and Embryology
University College London
Gower Street
London WC1E 6BT

INTRODUCTION

The essential property of teratocarcinoma stem cells is their pluripotency, that is their ability--like that of cells of an early embryo--to become diversely determined and to differentiate into a wide variety of cell types. Although clonal lines of these cells have been isolated and maintained in tissue culture, it has, until now, been possible to assay their pluripotency only by re-injection into a histocompatible host and attempts to obtain *in vitro* differentiation of these cells have met with limited success (reviewed by Evans, 1975 and Martin, 1975).

In contrast, it has been known for some time that embryoid bodies taken from the body cavity of a mouse carrying the ascites form of teratocarcinoma will differentiate extensively *in vitro* when they are cultured on a tissue culture substratum. Recently, it has been demonstrated (Teresky et al., 1974; Levine et al., 1974; Gearhart and Mintz, 1974) that this differentiation is dependent upon attachment of the embryoid bodies to the substratum and does not occur when they are kept in suspension. Lehman et al. (1974) have been able to establish a non-clonal cell line from embryoid bodies which produces a variety of differentiated cell types not present in

---

[1]Present address: Dept. of Pediatrics, University of California, San Francisco, California 94143.

the initial inoculum when the cells are maintained without subculturing for several weeks.

Studies of cell differentiation using embryoid bodies explanted from a mouse or non-clonal cultures have the disadvantage that the starting material is undefined. In order to examine the process of cell determination it is important to be able to initiate differentiation *in vitro* from homogeneous clonal cell lines. Previously, we (Martin and Evans, 1975a,b) have defined the conditions under which clonal teratocarcinoma cell cultures can be maintained in the undifferentiated state; we now describe in detail the culture conditions under which these clonal cell lines will differentiate *in vitro* to produce tissues which are characteristic of teratocarcinoma.

MATERIALS AND METHODS

Cell Culture

The cells were isolated and maintained as described previously (Martin and Evans, 1975a,b). (Specific methods of culture for cell differentiation are discussed in the text of this paper.) Dulbecco modified Eagle's medium + 10% calf serum was used throughout.

Cell Lines

The pluripotent cell lines studied can be grouped into two categories (Type II and III, see below) on the basis of their apparent differentiative capacity. The differences between these two groups are probably a result of differences in the amount of time they have been passaged in culture. They are all derived from a well-differentiated teratocarcinoma OTT 5568 produced by implantation of a 3 day embryo into an adult testis (Stevens, 1970). The mass cultures derived from this tumor were frozen as secondaries. Just prior to freezing, the clonal line, SIKR, was initiated (Evans, 1972). Ampoules of SIKR were frozen at the earliest opportunity and from one of these the main stock of SIKR which we previously described (Martin and Evans, 1974) was established. It was from this main stock that the sub-clones of Type II were isolated. Sub-clones of Type III were isolated from frozen stocks of both the secondary cultures of the tumor and from SIKR stocks frozen at the earliest time.

In order to be able to compare the properties of pluripotent and non-pluripotent embryonal carcinoma cell lines, we have isolated a clonal line of embryonal carcinoma cells from a tumor (LS 402C; Stevens, 1958) which does not differentiate *in vivo*. This 'nullipotent' cell line is a representative of Type I.

Histology

In order to facilitate histological procedures, the cells were grown in solvent-resistant tissue culture petri dishes (Permanox dishes, Lux Corporation, Thousand Oaks, California). The cell cultures were fixed with 2% glutaraldehyde in phosphate buffered saline dehydrated with alcohol and embedded in Spurr or Araldite plastic as formulated for use in electron microscopy. After polymerisation at 70°C, it was possible to peel the petri dish away from the plastic block, leaving the cells embedded. 8 μm sections were cut parallel to the plane of the petri dish using either a Jung K or a Sartorious sliding microtome. Sections were mounted, stained in toluidine blue in 1% sodium borate at 60°C, rinsed with water and 70% alcohol, dried and mounted in immersion oil or glycerol.

OBSERVATIONS AND RESULTS

In the preceding paper (Martin and Evans, 1975b), we have described the initiation and growth of cell cultures from a single isolated embryonal carcinoma cell. These clones grow progressively as a homogeneous population of embryonal carcinoma cells when maintained by serial transfer every 3-4 days on fibroblastic cell feeder layers. They will, however, undergo extensive differentiation when maintained under suitable culture conditions. Three such conditions are described.

Conditions for *In Vitro* Differentiation

Mass culture

When $0.8-4 \times 10^5$ cells/cm^2 are seeded on a tissue culture dish without added feeder cells, they grow initially as compact flat colonies of embryonal carcinoma cells. These become large, rounded and densely piled. As the culture grows it becomes very crowded. There is a considerable amount of cell death but subsequently a variety of differentiated cell types are observed surrounding and migrating out from dense central piles of cells. The first differentiated cells to appear are embryonic endoderm cells, a sheet of which migrates out to form a halo around the piled central colony (Fig. 1a). The next clearly identifiable type of differentiated cell to appear at the edge of the cell mass are neuronal cells which form large networks of cell processes (Fig. 1b). The ultrastructure of the processes confirms their identification as neuronal cells (Fig. 1c). By this time the central colony is becoming very densely multilayered and a complex of differentiated tissues is forming within it. The histology of these tissues is described below.

Clonal colony culture

When 50-100 cells are seeded on a confluent monolayer of feeder cells, clonal colonies form. They grow to a large size and by 4-5 weeks differentiated cells appear at the edge of the colony. By this time the feeder layer is degenerating and the piled centre of the colony is often necrotic. Endodermal cells appear as a migrating sheet and also invest marginal areas of embryonal carcinoma cells, (Fig. 2). Neuronal cells form networks of processes and some cell lines produce rhythmically beating muscle and areas of pigmented cells. Other tissues are formed within the multilayered central cell mass and are best revealed by histological sectioning.

From embryoid bodies formed *in vitro*

Embryoid bodies may be formed *in vitro* by homogeneous clonal embryonal carcinoma cell cultures as we have described elsewhere in this volume (Martin and Evans, 1975b). When the embryoid bodies, which are maintained in suspension in a nonadhesive bacteriological petri dish, are transferred to fresh medium and allowed to attach to an adhesive tissue-culture substratum, the endoderm migrates out as a halo around the central clump, followed by neuronal and other cell types. In subsequent growth and differentiation over the next 3-4 weeks, the culture becomes dense and multilayered and a wide variety of types of tissue develop.

Differentiative Behavior of Various Clones

The clones we have examined fall into 3 categories. Type 1 are non-pluripotent embryonal carcinoma cells--nullipotential cells--which do not differentiate either *in vivo* or *in vitro*. Aggregates of these cells do not develop any endoderm and the cells do not form embryoid bodies.

All the pluripotent cell lines (including some not described here) produce embryoid bodies *in vitro*. These consist of an outer layer of endodermal cells surrounding an inner core of embryonal carcinoma cells. These pluripotent lines fall into two categories on the basis of their subsequent behavior when they are maintained in suspension. Type II lines remain as simple embryoid bodies and there is no further differentiation. Only when these simple embryoid bodies are transferred to fresh medium and allowed to attach to an adhesive tissue-culture substratum does further differentiation take place.

Type III lines develop simple embryoid bodies in the same way as Type II lines and when these are replated on a tissue-culture substratum extensive differentiation also takes place. When the embryoid bodies formed by these cells are maintained in suspension, however, they undergo further differentiation to form cystic embryoid bodies. Subsequent differentiation within the embryoid body leads to a complex of different tissues. A description of this differentiation which has close analogies with embryonic development will be reported elsewhere (Martin and Evans, in preparation).

The differentiative behavior of these three categories of cell lines is diagramatically presented in Fig. 3. In each culture situation which leads to differentiation, the primary event is seen to be the formation of endoderm. As we have discussed (Martin and Evans, 1975b), it is when the embryonal carcinoma cells formed rounded aggregates that the endoderm forms on the free surface of these clumps. In the mass cultures (method 1) some clumps detach, form embryoid bodies in suspension and then re-attach, while other clumps form endoderm on their free surface without detaching from the substratum. In the clonal colony cultures (method 2) there is no detachment of the cells, but endoderm forms over the free surface of rounded masses of embryonal carcinoma. (Fig. 2 in this paper showing endoderm on a clonal colony is closely comparable with Fig. 6 of Martin and Evans (1975b), showing endoderm on an embryoid body in suspension). Nullipotential Type I lines will neither form embryoid bodies nor differentiate.

Differentiated Tissues Produced

As already indicated, differentiation of these cultures may be observed directly in the culture dish by phase contrast microscopy. The primary event in each case is the production of endodermal cells which may be recognized by their granular, sometimes vacuolated, cytoplasm and their refractive cell boundaries. Neuronal cells, fibroblastic cells, pigmented epithelial cells, rhythmically contracting muscle, cartilage and keratinising squamous epithelium which forms keratin pearls may all also be recognized in the living

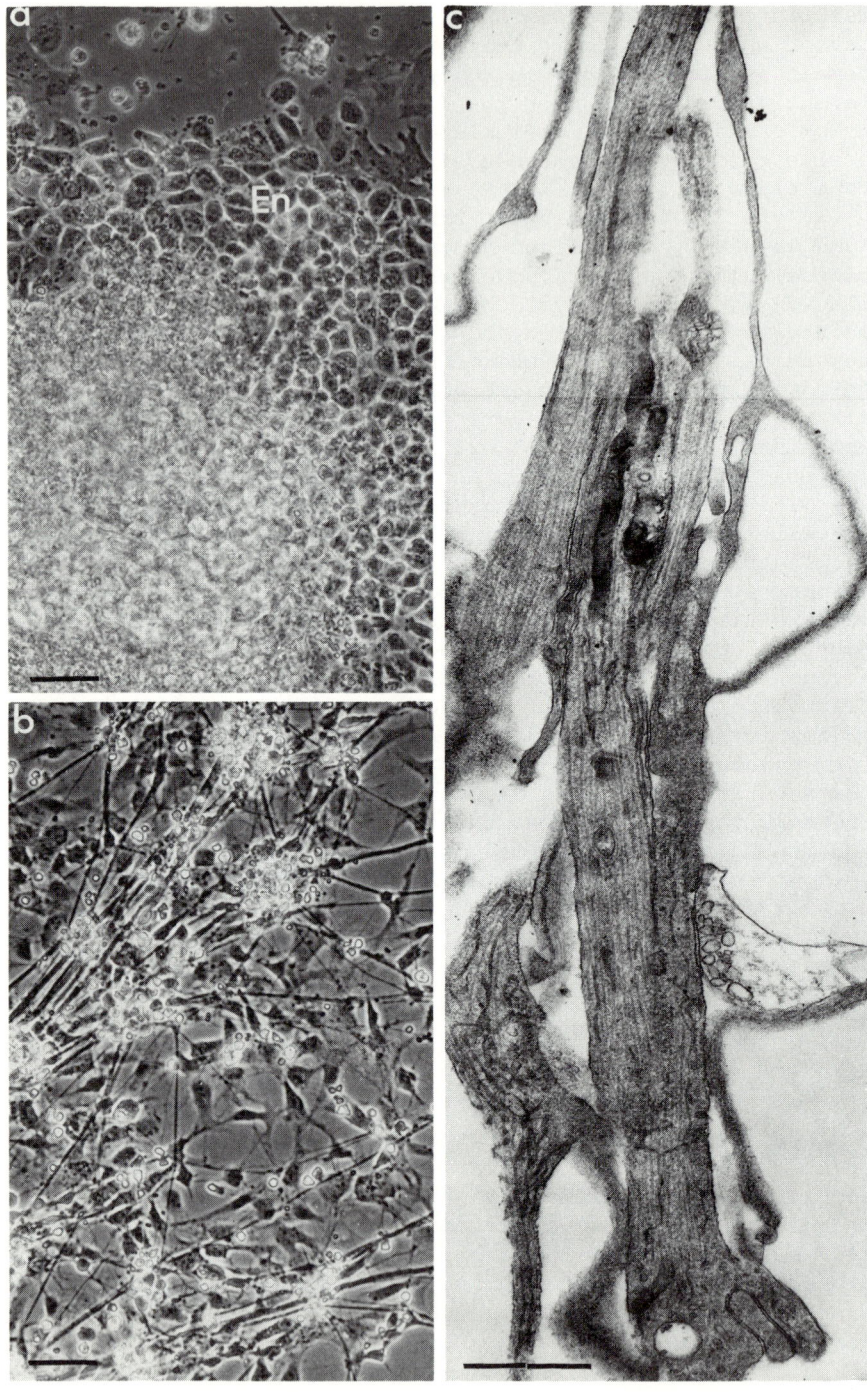

Figure 1. (Opposite page.)
a. Endodermal cells (En) migrating out from the central cell clump. Phase contrast microscopy. Scale bar = 50 um.
b. Neuronal cells. Phase contrast microscopy. Scale bar = 50 um.
c. Longitudinal section of a neuronal cell process. Electron micrograph. Scale bar = 1 um.

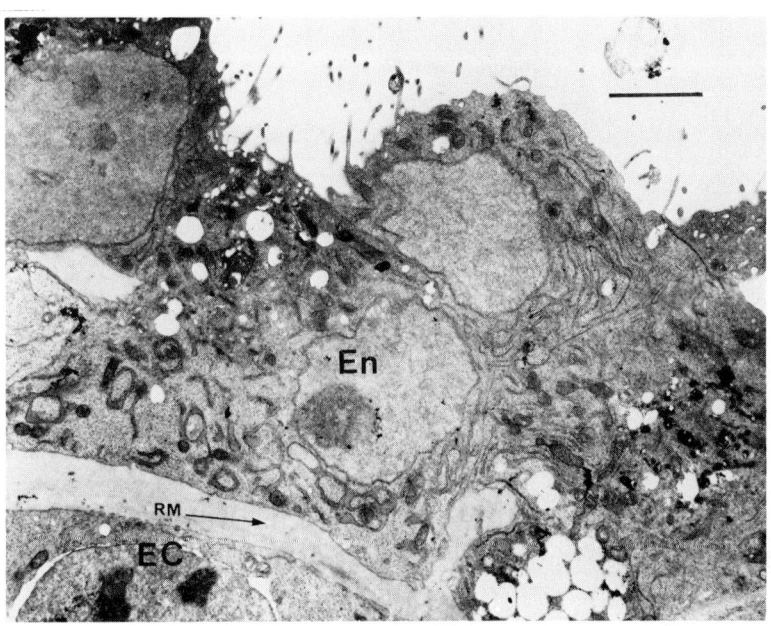

Figure 2. Electron micrograph of the cells at the edge of a differentiating clonal colony. Endodermal cells overlie embryonal carcinoma cells and are separated from them by Reichert's membrane-like material. En, endodermal cell; EC, embryonal carcinoma cell; RM, Reichert's membrane-like material. Scale bar = 1 um.

cultures. These tissues are, however, much better seen in histological sections of the differentiated cultures.

The cells are embedded in plastic on the petri dish and sections are cut in a plane parallel to the surface of the dish. These sections reveal a complex of tissues (Fig. 4) remarkably similar to that seen in histological sections from a teratoma produced by these cells *in vivo*. Embryonal carcinoma cells can be found in these sections so, formally, these cultures are the equivalent of teratocarcinomas. Derivatives of all three germ layers are found. Ectodermal derivatives

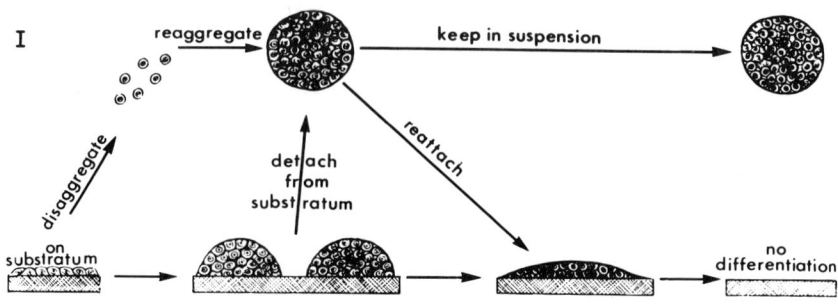
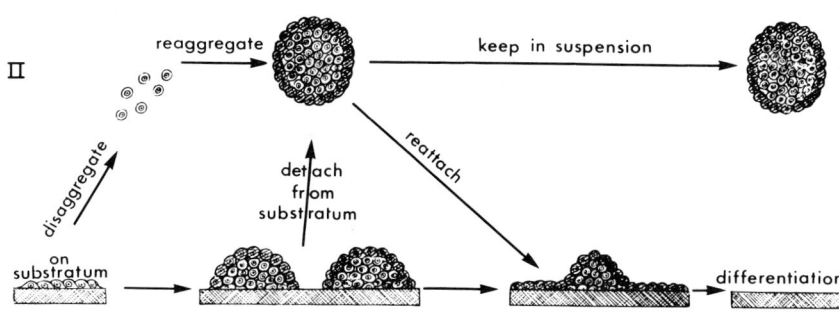
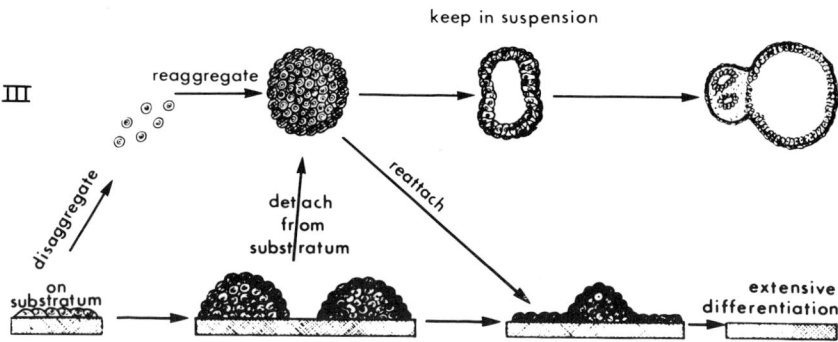

Figure 3. *Differentiative behavior of embryonal carcinoma lines, Types I-III.*

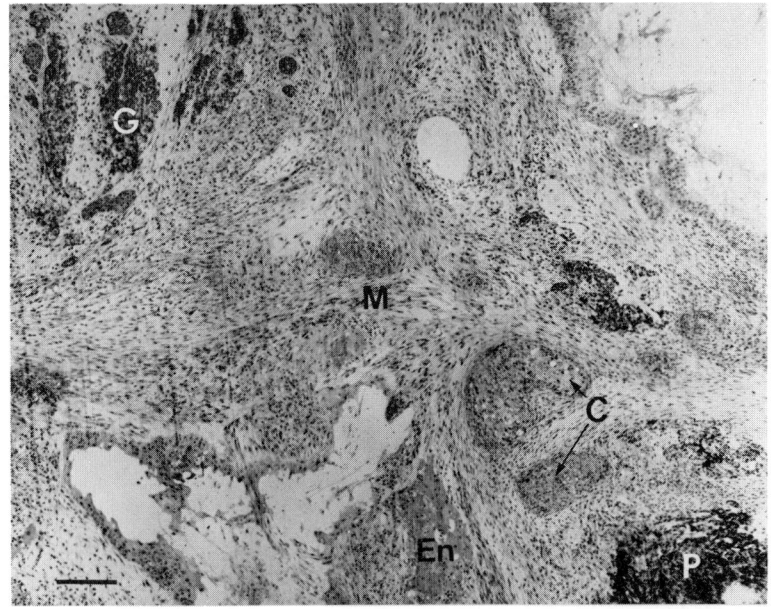

Figure 4. Histological section of a plastic-embedded differentiated cell culture. G, glandular tissue; M, mesenchymal tissue; C, cartilage; En, yolk sac endoderm; P, pigmented epithelium. Light microscopy. Scale bar = 250 µm.

are neural tissue, pigmented epithelium and keratinising squamous epithelium. Mesodermal derivatives include fibroblastic connective tissue, muscle and cartilage. Endodermal derivatives are represented by the embryonic endoderm and possibly by glandular structures although it is also possible that these are of ectodermal origin. Some examples of these tissues are shown in Fig. 5.

Although it is difficult to quantify, there appears to be some difference between the differentiation of the two categories of pluripotent line. Both *in vivo* and *in vitro* Type III lines (those which are able to form cystic embryoid bodies) produce a greater proportion of mature tissues and relatively less embryonal carcinoma, neuro-ependymal tubules and embryonic yolk sac endoderm than Type II. Even though the presence of a particular tissue is more significant than its apparent absence, beating muscle and pigmented epithelia which are very prominent in the living cultures are found only in the differentiated cultures from Type III (cystic) clones and not in those from Type II (simple).

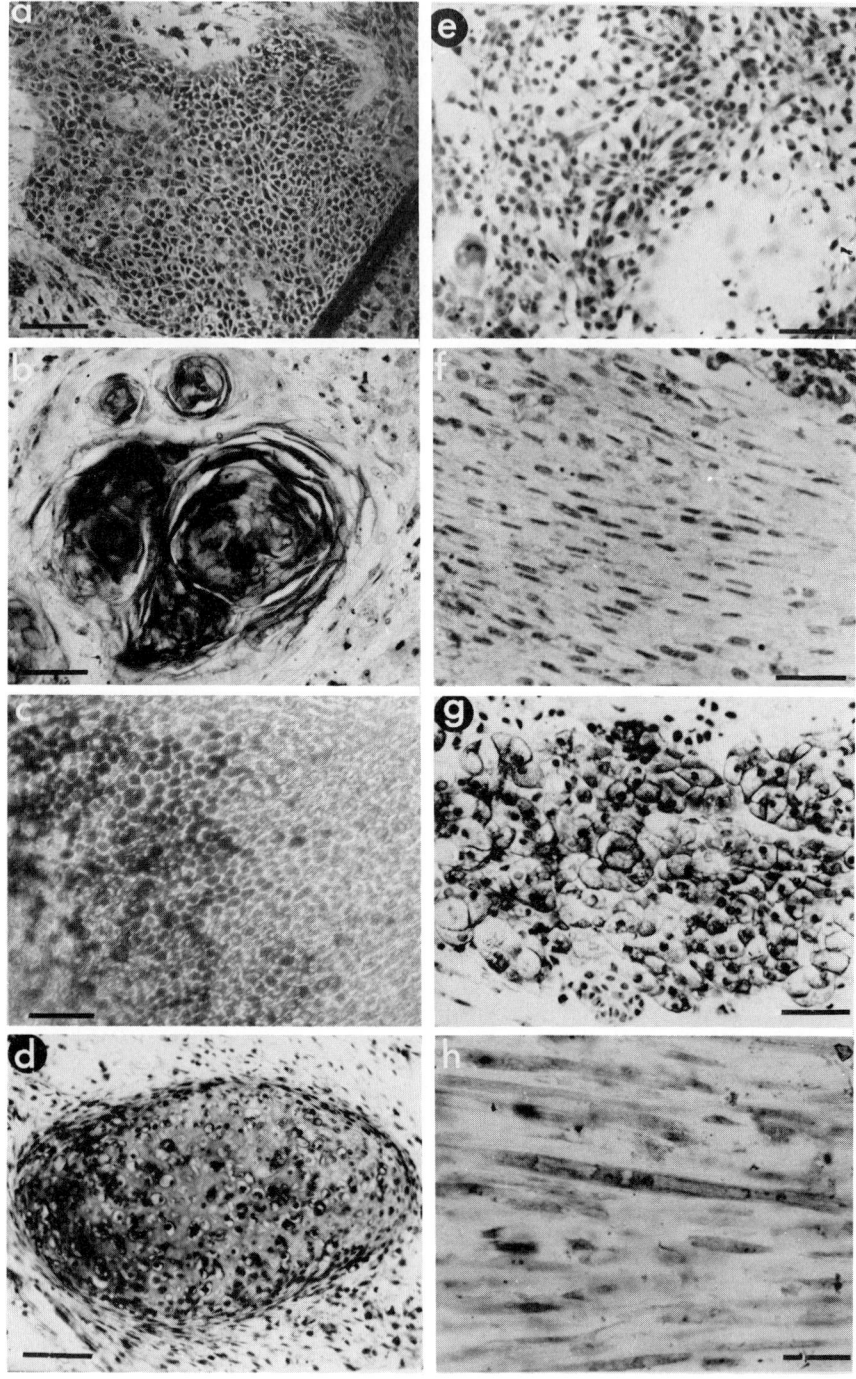

DISCUSSION

Teratocarcinoma stem cells are similar to the cells of early embryos in their ability to differentiate diversely. They are, moreover, tumor forming cells, but their differentiated progeny are usually non-tumorigenic. The ability to isolate clonal lines of these cells and to maintain them in the undifferentiated state in tissue culture makes it possible to explore the processes of cell determination and differentiation in a controlled environment which is accessible to detailed observation and experimental manipulation, providing that the differentiation of these clonal cell lines occurs *in vitro*.

Differentiation occurs when embryoid bodies are explanted from a mouse into tissue culture (Teresky et al., 1974; Levine et al., 1974; Gearhart and Mintz, 1974), but in this case the starting cell population is difficult to define and the contribution of the host environment to cell determination is unknown. Lehman et al. (1974) have reported the differentiation of embryonal carcinoma cells in cultures of a heterogeneous, non-clonal cell line derived from embryoid bodies. In this case the contribution of the host environment is eliminated, but the mixed cell population and their interactions are poorly defined. Although earlier attempts to obtain differentiation *in vitro* of clonal embryonal carcinoma cell cultures seemed to be unsuccessful, Jakob et al. (1973) noted that it did occur. We have here provided a detailed description of the differentiation of clonal teratocarcinoma cell lines *in vitro*.

Figure 5. Examples of various tissues formed in vitro.
a. Endodermal cells. Plastic-embedded section. Scale bar = 100 um.
b. Keratin pearls. Plastic-embedded section. Scale bar = 66 um.
c. Pigmented epithelium. Phase contrast microscopy. Scale bar = 55 um.
d. Cartilage. Plastic-embedded section. Scale bar = 100 um.
e. Neural tissue. Plastic embedded section. Scale bar = 100 um.
f. Mesenchymal tissue. Plastic embedded section. Scale bar = 100 um.
g. Glandular tissue. Plastic embedded section. Scale bar = 60 um.
h. Muscle fibers. Plastic embedded section. Scale bar = 125 um.

The differentiation observed *in vitro* is similar to that which occurs *in vivo*, both in the variety and arrangement of the tissues produced. This being the case, there are a number of advantages to be obtained in using the *in vitro* culture system as a means of assessing pluripotency. First, there is no requirement for a histocompatible host. Second, there is no possibility of host contribution to the differentiated tissues found; *in vivo*, the host contribution of nerves, blood vessels and mesenchymal stroma is ill-defined. Moreover, the differentiative ability of individual cells in a population can be assessed using the colony method (2) of obtaining differentiation described above.

In each situation in which these cells differentiate, the first event is the production of embryonic endoderm on the surface of a rounded group of cells. When these cells are in suspension, or if they become detached from the substratum, the resulting structure is a simple embryoid body. When the cell colony remains attached, the outer covering of endoderm which invests the embryonal carcinoma cells is in exactly the same relationship as in an embryoid body. These observations suggest that the formation of endoderm is a crucial step in the differentiation of teratocarcinoma stem cells and they explain why differentiation proceeds readily from embryoid bodies explanted from the animal.

As we have already discussed (Martin and Evans, 1975a,b), the formation of endoderm on the surface of a group of pluripotent cells parallels the early events of mouse embryogenesis. In other vertebrate embryos it is known that the endoderm has an inductive role in the organization of ectoderm development and the epigenesis of the mesoderm (Nieuwkoop, 1973; Waddington, 1952), and it is possible that the further differentiation of the teratocarcinoma stem cells is also dependent upon such interactions.

Embryoid bodies formed by Type II cultures remain simple and do not differentiate further, except when allowed to attach to a substratum. Type III cultures produce embryoid bodies which progress in suspension from simple to cystic, and in which further differentiation is not dependent upon attachment to a substratum. It is possible that the formation of a cavity in these cystic embryoid bodies provides an effectively two-dimensional surface for the positional organization of further cell determination and that simple embryoid bodies which have lost the ability to cavitate must spread onto a surface before this organization can proceed.

The observations that the first, and also the subsequent, differentiation of these cells as embryoid bodies parallels embryonic development (Martin and Evans, 1975a, 1975b) suggest that this differentiation provides a valid model for the study of embryonic development and does not represent an abberant 'neoplastic differentiation' (Lehman et al., 1974). The culture of these cells, therefore, provides an ideal system for the exploration of the events and processes of embryonic cell determination and differentiation. We have already examined the changes in expression of cell surface antigens during this process (Stern, Martin and Evans, in preparation). The first step--the production of endoderm--is clear and it is now possible to examine the control mechanisms and the biochemical processes of a defined determinative event. The influence of this endoderm on subsequent cell determination is also open to investigation.

ACKNOWLEDGEMENTS

We wish to thank Mrs. M. Reynolds, Mrs. P. Beveridge and Mrs. P. Pashley for their excellent technical assistance in the course of the work described here and in the manuscript by Martin and Evans in this volume.

This investigation was supported by a grant from the Cancer Research Campaign. GRM is supported by a fellowship from the U. S. Public Health Service, National Cancer Institute, 1 F22 CA04012-01.

REFERENCES

Evans, M. J. (1972). The isolation and properties of a clonal tissue culture strain of pluripotent mouse teratoma cells. J. Embryol. Exp. Morph. 28, 163.

Evans, M. J. (1975). Studies with teratoma cells *in vitro*. In "The Early Development of Mammals" (Eds. M. Balls and A. Wild). Cambridge University Press, London, in press.

Gearhart, J. D., Mintz, B. (1974). Contact-mediated myogenesis and increased acetylcholinesterase activity in primary cultures of mouse teratocarcinoma cells. Proc. Nat. Acad. Sci. U. S. 71, 1734.

Jakob, H., Boon, T., Gaillard, J., Nicolas, J.-F., Jacob, F. (1973). Teratocarcinome de la souris: isolement, culture et proprietes de cellules potentialites multiples. Ann. Microbiol. (Inst. Pasteur). 124B, 269.

Lehman, J. M., Speers, W. C., Swartzendruber, D. E., Pierce, G. B. (1974). Neoplastic differentiation: characteristics of cell lines derived from a murine teratocarcinoma. J. Cell Physiol. 84, 13.

Levine, A. J., Torosian, M., Sarokhan, A. J., Teresky, A. K. (1974). Biochemical criteria for the *in vitro* differentiation of embryoid bodies produced by a transplantable teratoma of mice. The production of acetylcholine esterase and creatine phosphokinase by teratoma cells. J. Cell Physiol. 84, 311.

Martin, G. R. (1975). Teratocarcinomas as a model system for the study of embryogenesis and neoplasia: a review. Cell 5, 229.

Martin, G. R., Evans, M. J. (1974). The morphology and growth of a pluripotent teratocarcinoma cell line and its derivatives in tissue culture. Cell 2, 163.

Martin, G. R., Evans, M. J. (1975a). The differentiation of clonal lines of teratocarcinoma cells: formation of embryoid bodies *in vitro*. Proc. Nat. Acad. Sci. U. S. 72, 1441.

Martin, G. R., Evans, M. J. (1975b). The formation of embryoid bodies *in vitro* by homogeneous embryonal carcinoma cell cultures derived from isolated single cells. This volume, p. 169.

Nieuwkoop, P. D. (1973). The "organisation centre" of the amphibian embryo: its origin, spatial organisation and morphogenetic action. Advan. Morph. 10, 1.

Stevens, L. C. (1968). Studies of transplantable testicular teratomas of strain 129 mice. J. Nat. Cancer Inst. 20, 1257.

Stevens, L. C. (1970). The development of transplantable teratocarcinomas from intratesticular grafts of pre- and postimplantation mouse embryos. Develop. Biol. 21, 364.

Teresky, A. K., Marsden, M., Kuff, E. L., Levine, A. J. (1974). Morphological criteria for the *in vitro* differentiation of embryoid bodies produced by a transplantable teratoma of mice. J. Cell Physiol. 84, 319.

Waddington, C. H. (1952). "The Epigenetics of Birds." Cambridge University Press, London.

THE *IN VITRO* DIFFERENTIATION OF EMBRYOID BODIES PRODUCED BY A TRANSPLANTABLE TERATOMA OF MICE

J. D. HALL, M. MARSDEN, D. RIFKIN[1],
A. K. TERESKY AND A. J. LEVINE

Princeton University
Department of Biochemical Sciences
Princeton, New Jersey   08540
and
[1]Rockefeller University
New York, New York   10021

INTRODUCTION

Testicular teratomas of mice represent a particularly useful system for the study of development and tumorigenesis (Stevens, 1967a,b; Pierce, 1967). In the 129/Sv strain of mice, these tumors apparently originate from male primordial germ cells during the twelfth to thirteenth day of fetal life (Stevens, 1962, 1967a,b). The available histological evidence indicates that the tumor gives rise to differentiated cells derived from all three germ layers (Pierce, 1967; Stevens, 1967a,b). These differentiated tissues appear normal by all histological criteria (Pierce, 1967).

Testicular teratomas can be serially transplanted in the 129/Sv inbred strain of mice (Stevens, 1958) producing tumors composed of two cell classes: (1) an embryonal carcinoma cell that is the pluripotent precursor or stem cell (Kleinsmith and Pierce, 1964; Kahan and Ephrussi, 1970; Rosenthal et al., 1970) or (2) a variety of differentiated nontumorigenic cell types (Pierce, 1967; Stevens, 1967a,b). A remarkable feature of these transplantable tumors is the dramatic influence of environmental factors on the tissue types produced. Subcutaneous injection of embryonal carcinoma cells results in a solid tumor that usually contains many well differentiated tissues (Pierce, 1967; Stevens, 1967a,b). These tissues, while disorganized when compared to a normally developing mouse embryo, are distributed in a clonal fashion so that identical cell types are clustered. Injection of the same embryonal carcinoma cells into the peritoneal cavity of the mouse results in spherical multicellular aggregates called embryoid bodies (Stevens, 1959; Pierce and Dixon, 1959). Simple embryoid bodies resemble normal five to six day old mouse embryos (Stevens, 1959; Teresky et al., 1974) with endodermal cells surrounding several layers of embryonal carcinoma cells (Teresky et al., 1974). A second cystic form of embryoid bodies contains mesodermal derivatives in addition to these two major cell types (Pierce and Dixon, 1959).

In order to study the events occurring during differentiation of testicular teratomas it appeared desirable to develop a system that permitted embryonal carcinoma cells, obtained directly from a transplantable tumor, to differentiate in cell culture. Accordingly, we have followed the growth and development of embryoid bodies *in vitro*, both in suspension culture and on the surface of a culture dish (Teresky et al., 1974; Levine et al., 1974). Three criteria have been employed to follow the differentiation of embryoid bodies *in vitro*: (1) morphological changes in the cell types produced in culture, (2) biochemical or enzymatic markers, characteristic of either specific differentiated tissues of the mouse or of transformed cells in culture, and (3) the ability of cells derived from embryoid body cultures to form tumors in 129/SvSl mice. The results of these studies (Teresky et al., 1974; Levine et al., 1974) indicate that many of the observations made with transplantable teratomas *in vivo* can be faithfully reproduced in cell culture.

As a further means of analysing this *in vitro* system, we have looked for antigens present on the surface of teratoma cells. Artzt et al. (1973; 1974) demonstrated the usefulness of such an approach when they detected a cell surface antigen on embryonal carcinoma cells and cleavage stage (morula) mouse embryos which appears to be specified or controlled by a locus affecting development, the T-locus. We have demonstrated that mouse serum from multiparous pregnant mice contains antibodies which cross react with cell surface antigens in cell lines derived from transplantable teratomas. These cell lines are composed of undifferentiated cells and a variety of differentiated cell types. In addition, these sera detect cell surface antigens on pure cloned cultures of embryonal carcinoma cells and on a parietal yolk sac carcinoma derived from a teratoma but not on normal adult fibroblasts in culture. Sera from pregnant mice, therefore, may be useful tools for detecting and isolating cell surface antigens common to normal fetal tissue and several teratoma cell types.

RESULTS

Morphological Changes in Primary Embryoid Body Cultures

Embryoid bodies from the OTT 6050A transplantable teratoma of 129/SvSl mice (Teresky et al., 1974) were purified by unit gravity sedimentation and placed in cell culture. These embryoid bodies are of the simple type, containing only embryonal carcinoma cells surrounded by a single layer of endodermal cells (Fig. la). Embryoid bodies were cultured under two conditions: (a) suspension growth (with agitation) or (b) growth on the surface of a culture dish. It was felt that suspension growth in culture might resemble growth in the ascites cavity where differentiation is limited to a few fetal tissue types. On the other hand, growth on the surface of the culture dish might be similar to that of solid teratomas where differentiation of stem cells is more extensive, resulting in the formation of a number of different mature tissue types.

When embryoid bodies composed of stem cells and endodermal cells (Teresky et al., 1974) are kept in suspension culture over a period of two to three weeks these structures undergo only limited morphological changes. A number of embryonic cell types appear including yolk sac membranes, mesoderm and, in some cases, structures resembling blood islands (Fig. lb, c, and d). In all cases, embryonal carcinoma cells are also observed. The pattern of development observed under these conditions resembles the development of a 5-6 day embryo into a 7-8 day one (Snell and Stevens, 1966). Longer culture periods in suspension do not lead to significant changes in the tissue types observed, even though these suspension cultures of embryoid bodies continue to increase in mass, cell number and DNA content (Teresky et al., 1974; Levine et al., 1974).

On the other hand, when these same embryoid bodies are grown on the surface of a culture dish, very different results are observed. During the first 5 to 10 days of culture growth, the embryoid bodies attach to the culture dish and clusters of cells surrounding these bodies are produced (Fig. 2), either by cell migration or proliferation. Stained preparations observed in the light microscope or the scanning electron microscope frequently reveal long fibrous projections of multinucleate cells. These multinucleate fibrous cells can cover entire areas of the culture dish (Fig. 2).

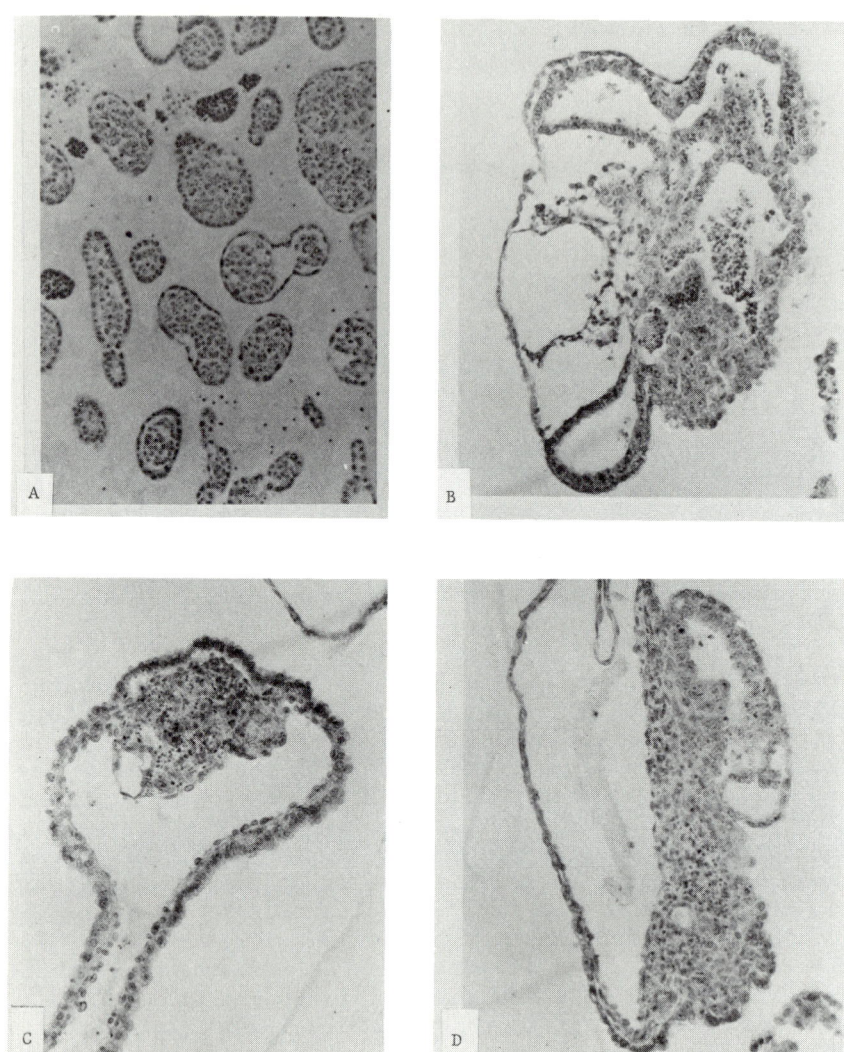

Figure 1. Light microscopy of embryoid bodies (OTT 6050A), sectioned and stained with eosin and hematoxylin. Embryoid bodies were obtained as described in Teresky et al. (1974). (a), embryoid bodies from the ascites cavity of a 129/SvSl mouse; (b), (c) and (d), embryoid bodies grown in suspension for 2 to 3 weeks.

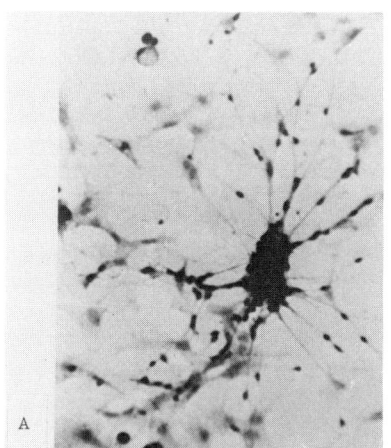
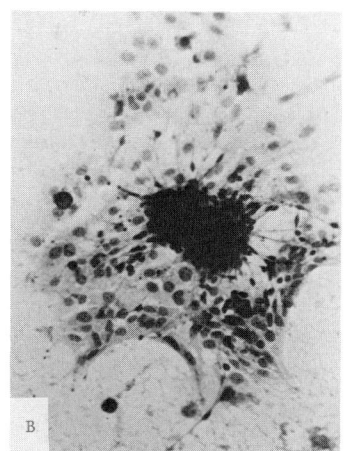
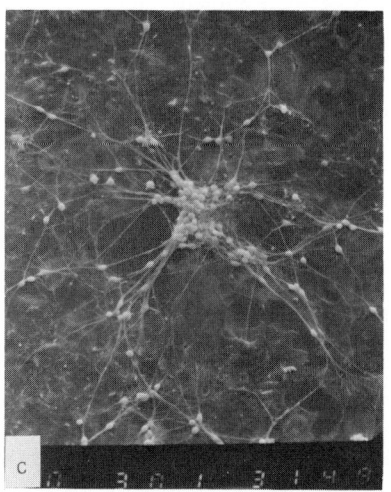
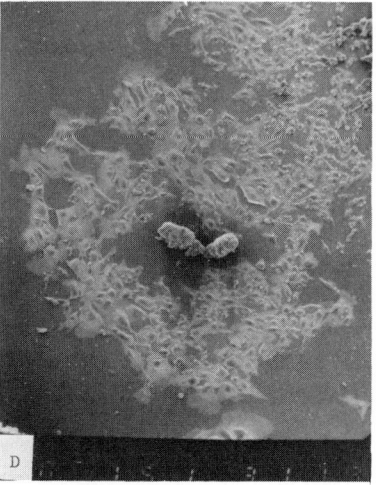

Figure 2. Embryoid bodies (OTT 6050A) growing on the surface of a culture dish for 7 days. (a) and (b), light microscopy of embryoid bodies and associated cells, fixed and stained with eosin and hematoxylin; (c) and (d), scanning electron micrographs of similar embryoid body cultures.

During the first two weeks in culture, there is a high rate of cell division, followed by the appearance of a number of new and diverse cell types (Fig. 3). By 3 to 4 weeks, the rate of cell division slows and the efficiency of plating or cloning falls as well (Levine et al., 1974). At early times in culture (1 week), between 35-60% of the cells will form colonies if diluted and replated, while by 3 to 4 weeks only 2-5% of the cells are able to form colonies under the same conditions.

The new cell types found after 3 to 4 weeks in culture are most frequently distributed on the culture dish so that cells of similar morphology are grouped together, e.g. Fig. 3. This clonal distribution of cell types could arise either by local proliferation of precursor cells or by migration of cells of similar morphology into a clonal distribution.

Based on these morphological studies, the growth of embryoid bodies *in vitro* shows definite similarities to embryoid body development *in vivo*. During suspension growth *in vitro*, the embryo-like structure of embryoid bodies is maintained and only limited, embryo-like, development is observed. In both these respects suspension growth resembles ascites growth *in vivo*. On the other hand, the growth of embryoid bodies on a petri dish surface resembles the development of teratoma cells in a solid tumor *in vivo*. In both cases, many differentiated cell types are observed to be arranged in a clonal fashion.

Biochemical Criteria for the Differentiation of Teratoma Cells in Culture

The morphological evidence presented above indicates that embryoid bodies in culture undergo changes characteristic of development and differentiation *in vivo*. It therefore seemed desirable to extend the criteria for differentiation *in vitro* to include biochemical markers characteristic of adult differentiated tissues and normal or transformed cells in culture. Because the transplantable teratoma employed in these studies (OTT 6050A) produced a large percentage of neuronal tissue types *in vivo* (Teresky et al., 1974; Levine et al., 1974) and neuronal-like cells *in vitro*, two enzymatic activities characteristic of mouse brain were chosen for study. Acetylcholine esterase activity is found predominantly in the brain and red blood cells of the mouse (Levine et al., 1974; Ellman et al., 1961). Creatine phosphokinase is predominantly found in brain and muscle tissue (Levine et al., 1974; Eppenberger et al., 1967). The brain and muscle forms of this enzyme activity are composed of different proteins which are readily distinguishable by gel electrophoresis (Levine et al., 1974). Transplantable teratomas of 129/SvSl mice composed of pure stem cells (402A) do not contain detect-

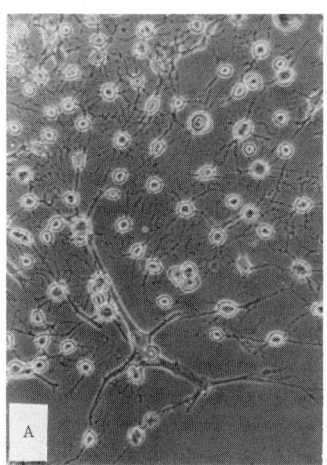
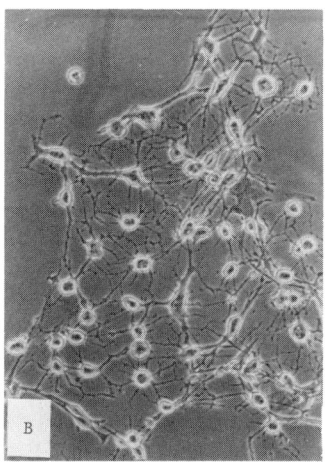
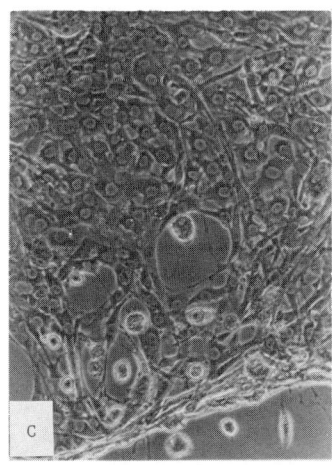
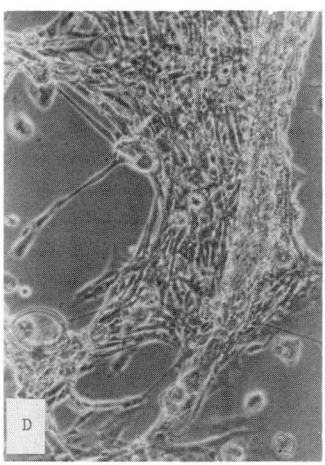

Figure 3. Different cell types derived in culture from embryoid body cells (OTT 6050A). (a) and (b) neuron-like cells; (c) epithelioid cells; (d) myoblast-like cells. Phase contrast. (From Teresky et al., 1974.)

able levels of either acetylcholine esterase or creatine phosphokinase, while well differentiated solid tumors (6050A) contain high levels of both enzyme activities (Levine et al., 1974). Hence, these enzyme activities indeed appear to be associated with differentiated tissue types of teratomas *in vivo* and may also be useful for following differentiation *in vitro*.

Embryoid bodies (OTT 6050A) from the ascites cavity of a 129/SvSl mouse have little or no acetylcholine esterase or creatine phosphokinase activity (Fig. 4, day 0). When these

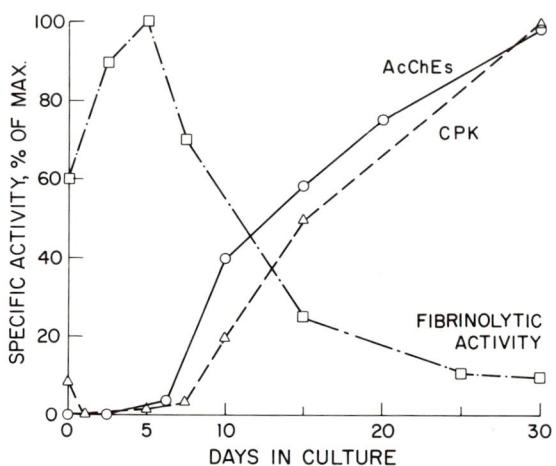

Figure 4. The kinetics of appearance of acetylcholine esterase (AcChEs), creatine phosphokinase (CPK) and fibrinolytic activity in cultures of embryoid bodies (OTT 6050A) grown on the surface of a petri dish for various times. The cell cultures and enzyme assays are described in Levine et al. (1974) and Unkeless et al. (1973).

structures were permitted to attach to the surface of a culture dish and allowed to differentiate, large increases in the specific activity of these two enzyme activities were observed (Fig. 4). Between day 5 and day 10 of culture growth, the specific activities of both enzymes increased above the background level and by day 30 specific activities had increased 70 to 100 fold (Fig. 4). Embryoid bodies kept in suspension culture over the same period of time did not show any significant levels of acetylcholine esterase or creatine phosphokinase activity (Levine et al., 1974). Thus, as in the case of solid teratomas *in vivo*, the appearance of differentiated cell types *in vitro* is correlated with an increase in these two enzymatic activities.

The acetylcholine esterase produced by these cells in culture hydrolyses acetylthiocholine 37-78 times more readily than butyrylthiocholine (Levine et al., 1974). The specificity of this enzyme activity found *in vitro* was identical to that observed for the activity found *in vivo* in differentiated solid teratomas. The great majority of the creatine phosphokinase activity found in the teratoma cells on the surface of a culture dish could be shown to be identical in its electrophoretic properties to the authentic mouse brain form of this enzyme (Levine et al., 1974). Differentiated solid teratomas also contain predominantly the brain form of creatine phosphokinase. Hence, these enzyme activities share similar properties both *in vivo* and *in vitro*.

Rous sarcoma virus transformed chick embryo fibroblasts and SV40 transformed rat embryo cells contain a proteolytic activity that cleaves plasminogen to plasmin. Normal cells do not have this protease activity (Unkeless et al., 1972; Ossowski et al., 1973a,b; Pollack et al., 1974). The presence of this protease has been correlated with transformation and can be assayed by the cleavage of plasminogen to plasmin. Plasmin is then detected by its ability to hydrolyze ^{125}I-labeled fibrin (Unkeless et al., 1973). Because embryonal carcinoma stem cells produce tumors in animals while differentiated cells are usually not capable of tumor formation, we investigated the levels of plasminogen-dependent fibrinolytic activity in embryoid bodies and cells differentiating in culture.

Embryoid bodies contained high levels of fibrinolytic activity (Table 1). This protease activity was dependent upon a source of plasminogen (calf serum) and was not detectable in plasminogen-free serum. Trypsin was employed as a positive control (100% hydrolysis) in these studies. Teratoma cells cultured for 25 days on the surface of the petri dish had only 13% of the protease activity found in embryoid bodies (Table 1). When embryoid body extracts were mixed with the extract from teratoma cells grown in culture for 25 days and the level of fibrinolytic activity tested, no indication of a soluble inhibitor present in the teratoma cell extract (25 days) was detected (Table 1). Hence, growth of teratoma cells in culture is associated with a reduction in this proteolytic activity.

The kinetics of cell associated fibrinolytic activity were examined as a function of time of embryoid body growth in cell culture (Fig. 4). An increase of fibrinolytic activity (between 2-10 fold) was consistently observed over a period of 5-7 days after embryoid bodies were placed in culture. The

TABLE 1

FIBRINOLYTIC ACTIVITY IN EXTRACTS OF EMBRYOID
BODIES AND TERATOMA CELLS IN CULTURE

		^{125}I-Fibrin - Cpm Hydrolysed		
Extract		0 Hrs	3 Hrs	6 Hrs
Embryoid Bodies, day 0	-	77	96	126
	calf	72	1066	1506
	plasminogen free-calf	75	82	101
Trypsin	-	245	1272	2028
-	-	52	60	74
-	calf	64	66	71
-	plasminogen free-calf	48	60	58
Cells in Culture, day 25	-	81	77	99
	calf	68	134	187
	plasminogen free-calf	60	48	55
Embroid Bodies, day 0 + Cells in Culture, day 25	-	66	81	100
	calf	79	930	1600
	plasminogen free-calf	75	100	110

Embryoid bodies and cell cultures were prepared as described in
Levine et al. (1974). Enzyme assays are described in Unkeless
et al. (1973).

increase was followed by a decline in activity to 13-22% of
the level observed for embryoid bodies (Table 1, Fig. 4).
This decline in activity was correlated with the appearance
of differentiated cell types as shown by increases in creatine phosphokinase and acetylcholine esterase activities.

The fibrinolytic activity in embryoid bodies may reside
in either embryonal carcinoma cells, endodermal cells or both.
The early increase in fibrinolytic activity after embryoid
bodies are placed in culture and attach to the surface of the
culture dish correlates well with the time of proliferation
of both embryonal carcinoma cells and endodermal cells. This
protease may, therefore, be a stem cell marker or alternatively it might be present on a cell type that is an intermediate
product of the stem cell during the process of differentiation.

Tumorigenic Potential of Differentiated Teratoma Cells in Culture

Based upon the morphological and biochemical evidence presented, embryoid bodies grown in suspension culture undergo only a limited developmental sequence, while cells attached to the surface of a culture dish show a number of differentiated properties. The available evidence with transplantable teratomas (Pierce, 1967) indicates that stem cells are tumorigenic while differentiated cells are usually not able to form a tumor in an animal. To determine if this correlation holds in embryoid body cell cultures the following experiment was performed. Embryoid bodies were placed in cell culture and cultured in suspension (with mild agitation) or on the surface of the culture dish (refed three times a week, passed 1/1 every 4 weeks) over a 12 week period. At appropriate times (Fig. 5) embryoid bodies were removed from the culture medium, or the cells were scraped from the surface of the dish with a rubber policeman. Approximately 10^4 cells (5 µg protein) from suspension culture and 10^4, 10^5 and 10^6 cells from the surface of the culture dish were injected subcutaneously into 129/SvSl mice. Tumor formation was examined over a one year period. Embryoid bodies grown in suspension culture over a 12 week period produced tumors in 100% of the mice. When approximately 10^4 cells grown on the surface of a petri dish were injected into 129/SvSl mice, only some of the mice developed tumors. Cells in culture for two weeks produced tumors in only 60% of the mice, while cells in culture for 8 weeks produced no tumors (0/5 mice) (Fig. 5). If 10^5 cells from the surface of a culture dish were injected into mice, no decrease in tumorigenic potential was observed until the cells had been cultured for at least 6 weeks and even after 12 weeks in culture this cell dose produced tumors in 50% (5/10 mice) of the animals injected. 10^6 cells from the surface of a culture dish produced 100% tumors in animals (5/5 mice) even after the cells had been grown in culture for 12 weeks before injection (Fig. 5).

The simplest interpretation of these data is that, during growth in culture, the cells found on the surface of the culture dish differentiate and are no longer able to form a tumor in an animal. The fact that tumors are produced with 10^5 cells (50%) and 10^6 cells (100%) but not with 10^4 cells (0%) after 12 weeks in culture indicates that the ratio of tumorigenic stem cells to nontumorigenic differentiated cells decreases with time in culture. Suspension growth favors limited development and the ratio of stem cells to differentiated cells is always high.

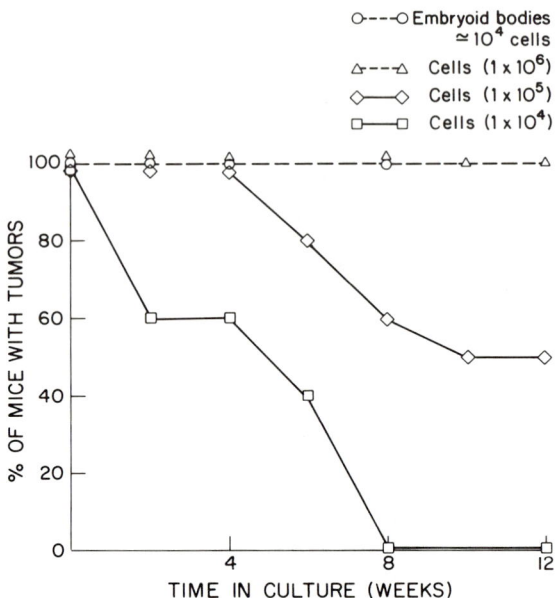

Figure 5. *Tumor-forming ability of cells from embryoid body (OTT 6050A) cultures grown for various times in suspension or on the surface of a culture dish. Cultures were prepared as described in Teresky et al. (1974). Cells were injected into 129/SvSl mice and the mice examined for tumors as described in the text. For injection of 10^5 cells per mouse, 10 mice were used for each time point. In all other cases, 5 mice were used for each point.*

Based upon the morphological, biochemical and tumorigenic properties to teratoma cells produced in culture we conclude that embryoid bodies attached to the surface of a culture dish differentiate into nontumorigenic cell types that produce several enzyme activities characteristic of adult tissue. Suspension growth in culture of these same embryoid bodies leads to more limited differentiation and distinctly fetal tissue types.

Fetal Cell Surface Antigens Found in Testicular Teratoma Cell Cultures

Artzt et al. (1973, 1974) have detected a cell surface antigen that is present on cleavage stage embryos and embryonal carcinoma cells. This antigen appears to be regulated or coded for by the T-locus of the mouse. Cell surface antigens might well be involved in regulating differentiation and morphogenic changes in the embryo (Bennett et al., 1972). As a first step in detecting and characterizing these cell surface antigens we have investigated the observation that multiparous pregnant mice have antibodies to teratoma cell surface antigens. Edidin et al. (1974) have recently shown that multiparous 129/J mice produce cytotoxic antibodies to teratocarcinoma cells in culture.

To detect antibody to teratocarcinoma cell surface antigens, test cells were grown in microwells (Linbro disposo trays). Cells were incubated for 2 hours at 32°C with dilutions of sera from pregnant or virgin female 129/SvSl mice. The cells were washed and incubated with ^{125}I-labeled rabbit anti-mouse gamma globulin Fab (Burdick and Wells, 1973). After 2 hours at 32°C the cells were washed again and the number of ^{125}I counts bound to the cells was determined in duplicate for each dilution of serum. The sera employed in this study all came from the 129/Sv inbred strain of mice. Pregnant mouse sera were usually obtained from multiparous mice (3 or more pregnancies) 15-20 days after mating. All sera were obtained and tested from individual mice (not pooled) by either tail bleeding or direct heart puncture. The virgin control sera (from two separate mice) came from females that were not age matched to the multiparous mice. Male 129/Sv mice yielded sera that did not contain detectable antibody to teratocarcinoma cell surface antigens (like virgin females).

Sera from pregnant mice contain antibody to cell surface antigens on cells from two independently derived cell lines of 129/SvSl testicular teratomas (OTT 6050A) (Fig. 6). 6050A-EB and 6050A-S are cell lines derived from embryoid bodies and a solid transplantable teratoma respectively. Both cell lines contain a mixture of differentiated cells and embryonal carcinoma cells by all the criteria reviewed in the previous section. These same preparations of sera from pregnant mice reacted with antigens on the cell surface of pure cloned lines of teratocarcinoma stem cells (PCC4·azal; Jakob et al., 1973) and a parietal yolk sac carcinoma derived from a teratoma (Fig. 7). However these sera did not detect cell surface antigens on secondary 129/SvSl baby mouse kidney fibroblasts (BMK) or an established 129/SvSl 3T3 (Postel and Levine, 1974) cell line (Fig. 8). Sera from mice in their first or second pregnancies contain little or no indication of these antibodies suggesting that multiple exposure to fetal tissue antigens is necessary for significant antibody production.

These data demonstrate that sera from multiparous pregnant mice contain antibody to cell surface antigens of testicular teratoma cell cultures, purified clones of embryonal carcinoma cells and a parietal yolk sac carcinoma cell culture. It appears likely that these sera from pregnant mice contain antibody to a number of different cell surface antigens and as such may be useful in isolating and characterizing cell surface changes occurring during development. Indeed, we have observed that sera from different multiparous pregnant mice may react differently on the same cell lines (Fig. 6, 7). The relative strength of different antibodies and even the presence of an antibody to a particular antigen may vary with the sera employed. The use of several independently derived sera from pregnant mice and cross absorption

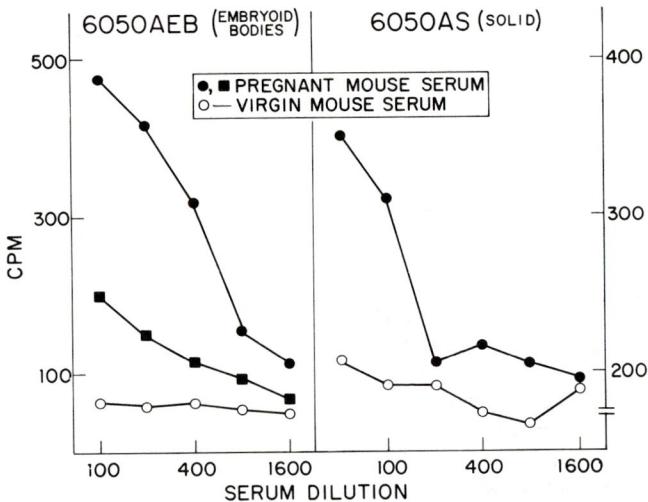

Figure 6. Binding of sera from pregnant or virgin 129/SvSl mice to cell lines derived from 129/SvSl teratoma OTT 6050A, 6050A-EB, embryoid body cultures, prepared as described in Teresky et al. (1974), and 6050A-S, cultures from solid teratomas, prepared from tumors resulting from subcutaneous injection of embryoid bodies into a 129/SvSl mouse. Serum binding was measured by the binding of ^{125}I-labeled rabbit anti-mouse immunoglobulin Fab to cells previously exposed to mouse serum. The ^{125}I-binding assay is described in Burdick and Wells (1973) and in the text. ● *and* ■ *represent serum obtained from two different multiparous pregnant mice.*

tests with the different positive cell lines should be useful in characterizing these different sets of antigens.

DISCUSSION

The experiments presented here demonstrate that growth of embryoid bodies in culture under the appropriate conditions leads to the development of differentiated cells *in vitro*. Three criteria have been employed to detect and characterize these developmental changes: 1) morphological alterations of cell types, 2) the appearance of enzyme activities characteristic of differentiated mouse tissue and 3) the decreased tumorigenic potential of differentiated cells.

It now remains to sort out the pathways and events involved in the processes of determination and differentiation.

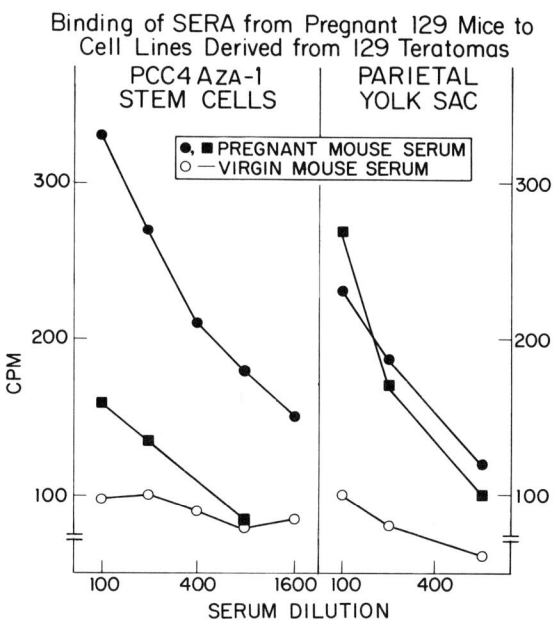

Figure 7. Binding of sera from pregnant or virgin 129/SvSl mice to cell lines derived from 129/SvSl teratomas, cultures of cloned PCC4.azal embryonal carcinoma cells (Jakob et al., 1973) derived from teratoma OTT 6050A and cultures of teratoma-derived parietal yolk sac carcinoma cells (provided by B. Pierce and J. Lehman). The legend to Figure 6 describes the binding assay.

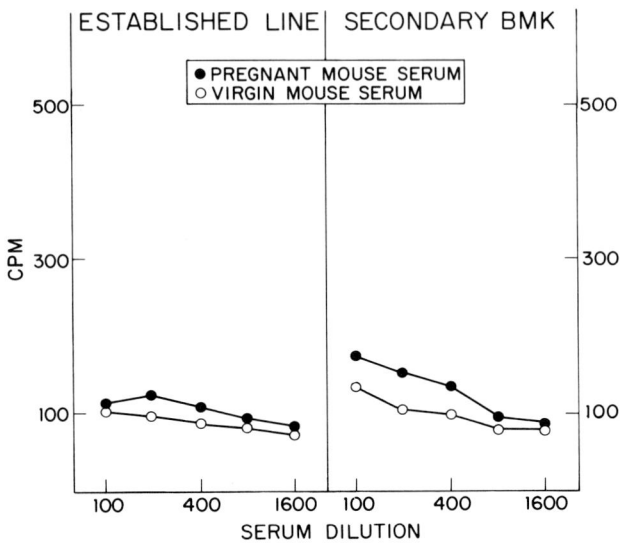

Figure 8. *Binding of sera from pregnant or virgin 129/SvSl mice to cell lines derived from 129/SvSl baby mouse kidney cells, an established 3T3 cell line and a secondary cell line. The legend to Figure 6 describes the binding assay.*

Several markers detected here may be useful for this purpose. The kinetics of appearance of the protease that cleaves plasminogen to plasmin indicates that it is present on or in stem cells or endodermal cells (or both) and therefore may be involved in the early events of embryoid body differentiation. Possibly, cell surface protease activities may participate in the reassortment of cells in a developing structure, the alteration of neighboring cell surface structure or the proliferation of cells in a local environment. As such, this protease could be a useful biochemical tool in the study of development. The plasminogen activator has been correlated with cellular transformation in Rous sarcoma virus infected chick embryo fibroblasts (Unkeless et al., 1973; Ossowski et al., 1973a,b) and in SV40 infected rat embryo cells (Pollack et al., 1974). The decreased levels of this protease activity which accompanies the loss of tumor forming ability of embryoid bodies during differentiation in culture strengthens this correlation in yet another system. The reciprocal relationship between differentiation and tumorigenesis in the teratoma system is one of the interesting facets of these studies (Pierce, 1967).

Clearly the environment that the stem cell finds itself in plays a critical role in determining gene expression in these cells. Both *in vitro* and *in vivo* suspension growth favors limited, fetal-like development and propagation of stem cells. Growth of a solid tumor or growth on a petri dish surface favors development of differentiated cell types and decreased proliferation of stem cells relative to differentiated cells. The molecular mechanism whereby these environmental factors affect stem cell differentiation is not clear at present. However, these results suggest that the cell surface and associated cellular structures (microfilaments, microtubules) may play a role in this process.

Sera from pregnant mice should be a most useful tool for seeking out cell surface markers produced during fetal or teratoma development. Embryonal carcinoma cells, parietal yolk sac cultures and teratoma cell cultures all contain surface antigens detected by antibodies in sera from multiparous pregnant mice, suggesting that these sera may contain antibodies to several different teratoma cell types. It is expected that sera from pregnant mice contain antibodies only to those fetal antigens that appear prior to the development of the immunological self tolerance system in the mouse. In this way adult tissue antigens should not elicit an antibody response within an inbred line of mice. Consistent with this assumption is the observation that secondary or established cultures of baby mouse kidney fibroblasts do not contain significant levels of cell surface antigens detected by the sera from pregnant mice. Further experiments are now required to elucidate the number and nature of these fetal cell surface antigens detected by the pregnant mouse serum. It is anticipated that this tool will be a useful one for understanding fetal development and teratoma differentiation.

ACKNOWLEDGEMENTS

We are indebted to the Denton Vacuum Company and the JEOL Corporation for the use of their equipment. The facilities of the Whitehall Foundation were generously supplied for these studies. The authors wish to acknowledge the help and advice of Dr. L. C. Stevens in this work and to thank Dr. N. Klinman for supplying the ^{125}I-anti mouse Fab serum. This research was supported by the National Science Foundation grant GB-38656.

REFERENCES

Artzt, K., Dubois, P., Bennett, D., Condamine, H., Babinet, C., Jacob, F. (1973). Surface antigens common to mouse cleavage embryos and primitive teratocarcinoma cells in culture. Proc. Nat. Acad. Sci. U. S. 70, 2988.

Artzt, K., Bennett, D., Jacob, F. (1974). Primitive teratocarcinoma cells express a differentiation antigen specified by a gene at the T-locus in the mouse. Proc. Nat. Acad. Sci. U. S. 71, 811.

Bennett, D., Boyse, E. A., Old, L. J. (1972). Cell surface immunogenetics in the study of morphogenesis. In "Proc. Third Lepetit Colloquium" (Ed. L. G. Silvestri) American Elsevier Publishing Co., New York, p. 247.

Burdick, J. F., Wells, S. A. (1973). Cross-reactivity between cell-surface antigens of different murine carcinogen-induced tumors, demonstrated by a modified isotopic antiglobulin test. J. Nat. Cancer Inst. 51, 1149.

Edidin, M., Gooding, L. R., Johnson, M. (1974). Surface antigens of normal early embryos and a tumor model system useful for their further study. Acta Endocrin. Suppl. 78, 336.

Ellman, G., Courtney, K., Andres, V., Featherstone, R. (1961). A new and rapid colorimetric determination of acetylcholine esterase activity. Biochem. Pharmacol. 7, 88.

Eppenberger, H. M., Dawson, D. M., Kaplan, N. O. (1967). The comparative enzymology of creatine kinase. I. Isolation and characterization from chicken and rabbit tissues. J. Biol. Chem. 242, 204.

Jakob, H., Boon, T., Gaillard, J., Nicolas, J.-F., Jacob, F. (1973). Teratocarcinome de la souris: isolement, culture et proprietes de cellules a potentialites multiples. Ann. Microbiol. (Inst. Pasteur) 124B, 269.

Kahan, B. W., Ephrussi, B. (1970). Developmental potentialities of clonal *in vitro* cultures of mouse testicular teratomas. J. Nat. Cancer Inst. 44, 1015.

Kleinsmith, L. J., Pierce, G. B., Jr. (1964). Multipotentiality of single embryonal carcinoma cells. Cancer Res. 24, 1544.

Levine, A. J., Torosian, M., Sarokhan, A. J., Teresky, A. K. (1974). Biochemical criteria for the *in vitro* differentiation of embryoid bodies produced by a transplantable teratoma of mice. The production of acetylcholine esterase and creatine phosphokinase by teratoma cells. J. Cell. Physiol. 84, 311.

Ossowski, L., Unkeless, J. C., Tobia, A., Quigley, J. P., Rifkin, D. B., Reich, E. (1973a). An enzymatic function associated with transformation of fibroblasts by oncogenic viruses. II. Mammalian fibroblast cultures transformed by DNA and RNA tumor viruses. J. Exp. Med. 137, 112.

Ossowski, L., Quigley, J., Kellerman, G., Reich, E. (1973b). Fibrinolysis associated with oncogenic transformation. J. Exp. Med. 138, 1056.

Pierce, G. B., Jr. (1967). Teratocarcinoma: model for a developmental concept of cancer. Curr. Topics Develop. Biol. 2, 223.

Pierce, G. B., Dixon, F. J. (1959a). Testicular teratomas. I. Demonstration of teratogenesis by metamorphosis of multipotential cells. Cancer 12, 573.

Pierce, G. B., Dixon, F. D. (1959b). Testicular teratomas. II. Teratocarcinoma as an ascitic tumor. Cancer 12, 584.

Pollack, R., Risser, R., Colon, S., Rifkin, D. (1974). Plasminogen activator production accompanies loss of anchorage regulation in transformation of primary rat embryo cell by simian virus 40. Proc. Nat. Acad. Sci. 71, 4792.

Postel, E. H., Levine, A. J. (1974). Studies on the regulation of deoxypyrimidine kinases in normal, SV40-transformed and SV40- and Adenovirus-infected mouse cells in culture. Virology 63, 404.

Rosenthal, M. D., Wishnow, R. M., Sato, G. H. (1970). *In vitro* growth and differentiation of clonal populations of multipotential mouse cells derived from a transplantable testicular teratocarcinoma. J. Nat. Cancer Inst. 44, 1001.

Snell, G. D., Stevens, L. C. (1966). Early embryology. In "Biology of the Laboratory Mouse" (Ed. E. Green) McGraw-Hill, New York, p. 205.

Stevens, L. C. (1958). Studies on transplantable testicular teratomas of strain 129 mice. J. Nat. Cancer Inst. 20, 1257.

Stevens, L. C. (1959). Embryology of testicular teratomas in strain 129 mice. J. Nat. Cancer Inst. 23, 1249.

Stevens, L. C. (1962). Testicular teratomas in fetal mice. J. Nat. Cancer Inst. 28, 247.

Stevens, L. C. (1967a). Origin of testicular teratomas from primordial germ cells in mice. J. Nat. Cancer Inst. 38, 549.

Stevens, L. C. (1967b). The biology of teratomas. Advan. Morph. 6, 1.

Teresky, A. K., Marsden, M., Kuff, E. L., Levine, A. J. (1974). Morphological criteria for the *in vitro* differentiation of embryoid bodies produced by a transplantable teratoma of mice. J. Cell. Physiol. 84, 319.

Unkeless, J. C., Tobia, A., Ossowski, L., Quigley, J. P., Rifkin, D. B., Reich, E. (1973). An enzymatic function associated with transformation of fibroblasts by oncogenic viruses. I. Chick embryo fibroblast cultures transformed by avian RNA tumor viruses. J. Exp. Med. 137, 85.

ALKALINE PHOSPHATASE ACTIVITY IN EMBRYONAL CARCINOMA AND ITS HYBRIDS WITH NEUROBLASTOMA

EDWARD G. BERNSTINE[1] AND BORIS EPHRUSSI[2]

[1]Biology Division, Oak Ridge National Laboratory,
Oak Ridge, Tennessee 37830
and
[2]Centre de Genetique Moleculaire du C.N.R.S.,
91190 - Gif-sur-Yvette, France

INTRODUCTION

The pattern of alkaline phosphatase activity in the developing mouse embryo is an intriguing phenomenon. The detailed studies of Mulnard (1955) have shown that the enzyme is first detectable by histochemical methods at the morula stage. Activity increases in the implanting blastocyst where it is localized in the inner cell mass. At the time of germ layer formation APase is essentially totally localized in the ectoderm. Thereafter, the distribution of activity is complex, high levels being found in virtually all organ primordia (see Moog, 1965) at early stages of their differentiation. This activity decreases markedly as cell differentiation and organogenesis progress. Thus, elevated enzyme activity is present in tissues differentiating from all three primary germ layers and is in some way abolished or greatly diminished during the final stages of differentiation.

A general role for APase in cell differentiation is implied by these observations as is the existence of a regulatory mechanism that controls the drastic changes in APase levels during embryogenesis. Neither the physiological function nor the means of regulating APase activity in embryos is currently understood. The results described in this report indicate that murine teratocarcinomas may provide favorable material for investigating these embryological problems.

ALKALINE PHOSPHATASE IN EMBRYONAL CARCINOMA

Damjanov et al. (1971) obtained histochemical evidence that strong APase activity in primary growths of seven-day mouse embryos grafted beneath the kidney capsule is found in regions undergoing ossification and in foci of cells that were identified as embryonal carcinoma. Our work on APase in tumors and cell lines derived from transplantable teratocinomas has been performed primarily using material obtained from the embryo-derived tumor OTT 6050 (Stevens, 1970). Details of experimental methods employed have been described previously (Bernstine et al., 1973, 1975).

The results of APase assays performed on extracts of tumors and embryoid bodies known to differ in their content of embryonal carcinoma provided us with a first indication that a positive correlation might exist between APase specific activity and the embryonal carcinoma content of teratocarcinomas *in vivo*. Representative data are given in Table 1. The lowest values for APase specific activity obtained were those for OTT 6050-943, a line that consists only of malignant endodermal cells. Neither embryonal carcinoma nor any differentiated cell types other than parietal endoderm are found in tumors produced by this line. Tumors of lines OTT 6050 B and OTT 6050-970 contain a mixture of embryonal carcinoma and a wide variety of differentiated cell types. The specific activity of APase in these tumors is intermediate between that of OTT 6050-943 and 402 A, which consists uniquely of embryonal carcinoma cells and has the highest level of APase activity.

Extracts of two types of embryoid bodies (Pierce, 1967) were also assayed for APase. Those from OTT 6050-970 are large cystic structures that often contain mesodermal derivatives and a relatively low proportion of embryonal carcinoma. Those from OTT 6050 B are smaller, consisting of a core of embryonal carcinoma surrounded by a layer of endoderm. As

---

ABBREVIATIONS AND TERMINOLOGY

APase - alkaline phosphatase (EC 3.1.3.1).
SDS - sodium dodecylsulfate.
AChE - acetylcholinesterase (EC 3.1.1.7).
NT hybrids - hybrid cells resulting from the fusion of a clonal line of embryonal carcinoma (H3.6) with a clonal line of neuroblastoma (N18TG2).
s - modal number of chromosomes.

TABLE 1

SPECIFIC ACTIVITY[a,b] OF APase IN TERATOMA LINES *IN VITRO*

Teratoma Line	Tumors	Embryoid Bodies
OTT 6050-970	2.3	4.2
OTT 6050 B	3.3	16.7
OTT 6050-943	0.8	1.2[c]
402 A	12.9	[d]

[a]Data are from Bernstine et al. (1973).

[b]μmol of p-nitrophenol released from p-nitrophenyl phosphate at pH 10 and 37°C per hour per mg of protein. Each value is the mean of 2 independent determinations.

[c]Intraperitoneal inoculation of OTT 6050-943 gives, in addition to tumors, free-floating clusters of parietal endoderm cells that are not true embryoid bodies. They contain no embryonal carcinoma.

[d]402 A, which is generally transplanted subcutaneously, produces no embryoid bodies.

shown in Table 1, APase specific activity in the small, simple embryoid bodies, rich in embryonal carcinoma, is four times that of the large, cystic structures. The histochemical staining shown in Figure 1 demonstrates that APase activity is confined to embryonal carcinoma in both types of embryoid body.

The APase activities of tissue culture cell lines isolated from embryoid bodies and classified as either embryonal carcinoma or non-embryonal carcinoma on the basis of cell morphology, tumorigenicity and tumor morphology are given in Table 2. The data show that APase is expressed in embryonal carcinoma cell lines *in vitro* where its specific activity may be as much as one hundred times that of non-embryonal carcinoma lines.

Since various isoenzymes of APase have been identified (see Fishman, 1974 for a review), it is of interest to determine whether the APase expressed in embryonal carcinoma is also present in adult tissues. The embryonal carcinoma enzyme can be distinguished from those of adult intestine and liver of mice of strain 129/Sv-Sl on the basis of its heat denaturation kinetics and the extent to which it is inhibited by L-phenylalanine (Bernstine et al., 1973). No differences between the teratoma APase and the enzymes from kidney and placenta were found by these methods. Figure 2 shows that the APase activities of embryoid bodies, placenta and kidney have different elution patterns from columns of DEAE-cellulose. Neither of the two peaks of embryoid body APase activity coincides with the peaks observed for the kidney and placental enzymes. It should be emphasized that these chromatographic differences must be considered as only a preliminary indication that differences exist among these enzymes. In each case, the APase applied to the columns was only twenty-fold purified over crude homogenates. Thus, the possibility that the enzyme is associated with other cellular

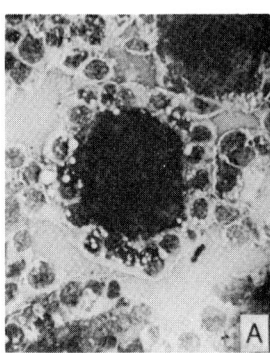 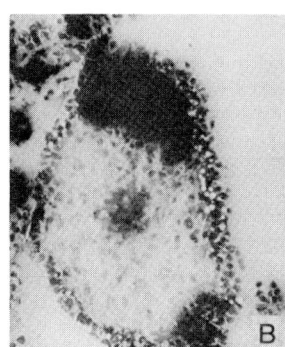

Figure 1. Histochemical localization of APase in embryoid bodies. (A) Embryoid body from OTT 6050 B with a core of embryonal carcinoma cells showing intense activity. Note absence of activity in the endoderm. (B) Large, cystic embryoid body from OTT 6050-970. Only nodules of embryonal carcinoma give a positive reaction, while mesenchyme and surrounding endoderm are negative. Nuclei in both preparations are counterstained with hematoxylin. From Bernstine et al., 1973.

components or otherwise aggregated must be considered. Both peaks of embryoid body APase maintain their position of elution upon re-chromatography and are inhibited to the same extent by L-phenylalanine.

ALKALINE PHOSPHATASE IN HYBRIDS BETWEEN EMBRYONAL CARCINOMA AND NEUROBLASTOMA

Cell fusion experiments in a variety of systems have shown that, as a rule, biochemical functions characteristic of specialized cell lines are not expressed in hybrid cells (see

TABLE 2

SPECIFIC ACTIVITY OF APase IN TERATOMA CELL LINES *IN VITRO*

Cell Line[a,b]	Cell Type	Specific Activity of APase (U/mg)
F9	Embryonal carcinoma	17.0
H3.6	"	34.1
H3.2	"	10.5
H3.4	"	15.4
H3.5	"	10.0
PC.2	"	19.8
PC.7	"	19.3
PC.8	"	28.7
PC.13	"	18.4
A2	"	3.4
A3	"	3.1
970-W	Non-embryonal carcinoma	0.7
970-Oa4	"	0.3
970-02C	"	0.6
970-P	"	<0.2

[a] Data are from Bernstine et al. (1973).

[b] Not included in the table are data on PPLO-free derivatives of H3.6 and F9.

Ephrussi, 1972 and Davidson, 1974 for reviews and complete references). Since the high level of APase activity in embryonal carcinoma is a biochemical characteristic of this cell type, it was of interest to examine the expression of this enzyme in hybrid cells. Crosses were made between embryonal carcinoma line H3.6 and a clonal line of mouse neuroblastoma marked with resistance to 6-thioguanine (Minna et al., 1972) that was kindly given to us by John Minna.

Figure 3 represents the lineage of the three original independent colonies of hybrid cells isolated, and clones and subclones derived from them. NT1 and NT2 exhibit marked morphological heterogeneity, whereas NT3 is a homogeneous population of fibroblastic cells. The morphological heterogeneity observed in the early generations of hybrid clones was much less evident in subclones derived later. Each of the subclones of NT1B, for example, is a relatively homogeneous population. Morphological distinctions between subclones are evident, some being reminiscent of parental lines, while others appear to have completely 'new' morphologies in culture. The subclones of NT1B have narrower distributions of total chromosomes than other hybrid lines (Table 3).

The karyological data presented in Table 3 support the view that the NT hybrids arose from the fusion of 1s embryonal carcinoma with 1s neuroblastoma cells. All chromosome

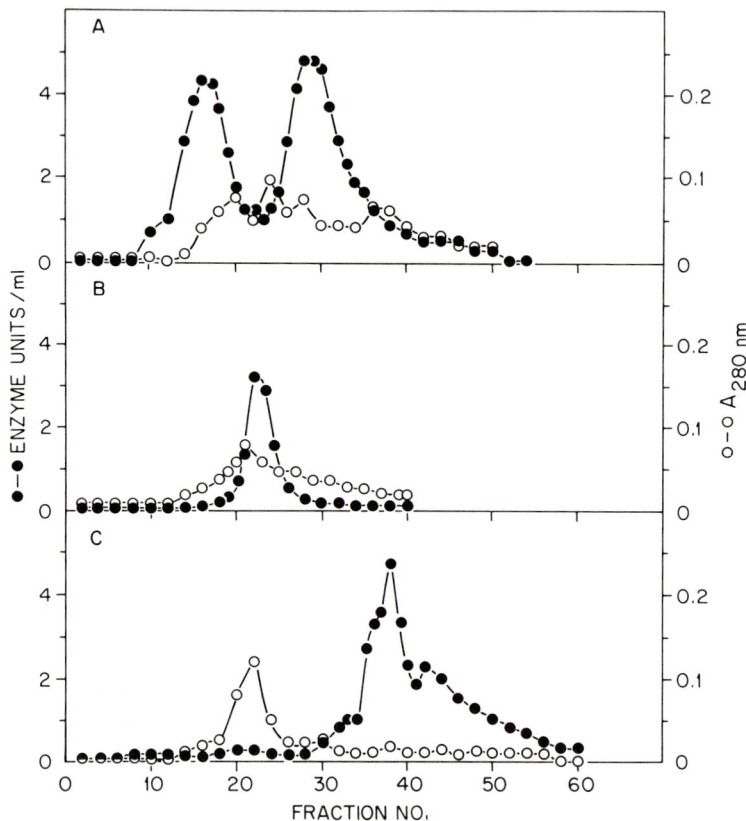

Figure 2. DEAE-cellulose chromatography of APases. APases from (A) embryoid bodies, (B) placenta and (C) kidney of 129/Sv-S1 mice were partially purified by butanol extraction (Morton, 1953), acetone precipitation and $(NH_4)_2SO_4$ fractionation and applied to columns (1 x 20 cm) of DEAE-cellulose equilibrated with 5 x 10^{-3} M Tris-HCl pH 7.2 containing 10^{-3} M $MgCl_2$. Yields were 60 to 80% of the activity present in crude homogenates. After washing with the equilibration buffer, chromatograms were developed with linear gradients of KCl (0-0.5 M) in the same buffer. ●—●, APase activity; o—o, absorbancy at 280 nm.

preparations from clone NT1F and four of twenty-one metaphases from clone NT2B contained unusually large numbers of chromosomes. Since all other clones derived from NT1 have chromosomal complements consistent with a *1s + 1s* origin, NT1F may have arisen either from a secondary fusion event or as a mitotic anomaly in the original hybrid colony. A high frequency of mitotic anomalies or chromosomal rearrangements in early hybrid generations is indicated by the fact that the clones derived from NT1 possess more biarmed chromosomes than can be accounted for by summing the number of such chromosomes in the parental lines.

Dev et al. (1973) have shown that two pairs of autosomes

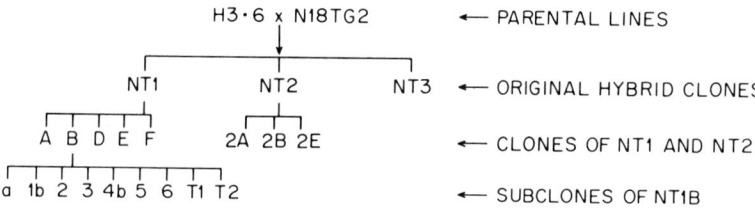

Figure 3. Relationships among hybrid cell lines derived from the fusion of H3.6 with N18TG2.

of 129 mice (Nos. 8 and 14) lack centromeric heterochromatin. Metaphases of H3.6 cells contain one large metacentric chromosome, devoid of centromeric heterochromatin (Fig. 4), which appears to be the product of centric fusion of two C-band deficient telocentrics. Since all biarmed chromosomes of neuroblastoma origin possess large blocks of centromeric heterochromatin (Fig. 4), we were able to identify the H3.6 metacentric in derivatives of all hybrid clones. This unambiguously shows that the NT hybrids are, indeed, hybrids.

Inoculations of derivatives of all hybrid clones except NT3 into (129/Sv-Sl x A/HeOrl)F1 animals have given tumors. A total of twenty-three tumors arising from the inoculation of thirteen hybrid clones has been examined histologically. We have observed no organized structures in any of these tumors, each of which appears to be composed of one predominant cell type; differences between tumors produced by various clones have been noted. The results of APase assays performed on stationary-phase cultures of parental and hybrid cell lines are shown in Table 4. The high level of enzyme activity in the hybrids is striking. Specific activities ranging from two to eight times that of H3.6 are found among

TABLE 3

CHROMOSOME ANALYSIS OF PARENTAL AND HYBRID CELL LINES

Cell Line	Total Number of Chromosomes		Number of Large Metacentrics		Number of Metaphases with	
	Range	Mode	Range	Mode	One Very Small Metacentric	One Submetacentric
H3.6 (21)[a]	38-46	42	—[d]	1	0	0
N18TG2 (31)	79-86[b]	79	—[d]	2	0	0
NT1 (13)	84-97	90	6-9	7	30	26[f]
NT1A (8)	97-119	109	4-10	6,7	12	9
NT1B (19)	102-146	N.C.M.[c]	6-13	N.C.M.[c]	8	6
1a (17)	96-116	101,102	5-10	6,7	19	18
1b (19)	100-103	101,102	7-9	7	17	16
4b (10)	93-103	99	4-8	6	0	18
5 (10)	98-105	98,102	5-7	7	10	10
6 (10)	101-104	103	6-8	6	10	10
T1 (10)	99-107	104	5-8	7	10	1
T2 (17)	103-111	N.C.M.[c]	6-8	8	10	10
NT1D (12)	94-109	102,104	5-8	7	16	16
NT1E (11)	91-106	103	8-12	9	10	10
NT1F (18)	104-126	N.C.M.[c]	5-12	8	9	8
NT2A (20)	147-274	N.C.M.[c]	N.D.[e]	N.D.[e]	N.D.[e]	N.D.[e]
NT2B (21)	88-113	109	6-11	7	14	18
NT2E (15)	107-232	119,122	N.D.[e]	N.D.[e]	N.D.[e]	N.D.[e]
NT3 (11)	101-131	111	7-12	9	15	12
	80-105	N.C.M.[c]	3-8	5	119	11

[a]Numbers in parentheses represent the number of metaphases examined.
[b]About 50% of H3.6 metaphases were 2s.
[c]No clear mode.
[d]All metaphases examined had the indicated number of large metacentric chromosomes.
[e]Not determined.
[f]Three of these had two submetacentric chromosomes.
[g]One of these had two very small metacentrics.

the hybrids. The large errors associated with the specific activities given in the table reflect slight differences in the growth phases of cultures used in different experiments (see below). The possibility that there is some heterogeneity in the level of APase expression within each of these cell lines has not been ruled out (see below). Heat denaturation studies (Fig. 5 and Bernstine et al., 1975) are consistent with the hypothesis that it is the embryonal carcinoma APase that is being expressed in the NT hybrids. Assay of APase in mixed extracts of N18TG2 and H3.6 gave no evidence for the presence of a diffusible activator.

The level of APase activity of embryonal carcinoma cells increases by about a factor of two in the stationary phase (Bernstine et al., 1973). Similar results were obtained for N18TG2 (Fig. 6). Unlike the parental lines, NT1 and NT1A (the only lines examined so far for this effect) show three- to eight-fold increases in specific activity after the cessation of cell division. In the logarithmic phase of the growth cycle the specific activity of APase in NT1 is *lower* than that of H3.6, whereas APase activity in NT1A is about equal to that of H3.6. Both hybrid lines attain higher levels of APase than H3.6 in the stationary phase. It will be of interest to measure the APase activity of the other hybrid lines as a function of time in culture in order to determine whether the enhancement of activity with respect to H3.6 reflects in all cases large increases during stationary phase.

Segregation for the expression of APase was found among subclones derived from NT1B (Table 5). The seven subclones assayed for APase have specific activities ranging from about four times greater to over two orders of magnitude less than that of H3.6. These results taken with the narrower distribution of chromosome numbers within each subclone of NT1B compared to the other hybrid lines (Table 3) suggest that the latter will also be found to be heterogeneous for APase expression. Thus, specific activities even higher than those given in Table 4 are expected among some subclones of these lines.

DISCUSSION

A high level of APase expression in embryonal carcinoma both *in vivo* and *in vitro* has been demonstrated. Furthermore, it is clear from our work and that of Damjanov et al. (1971) that most cell types found in teratocarcinomas, all of which differentiate from embryonal carcinoma (Kleinsmith and

Pierce, 1964; Kahan and Ephrussi, 1970; Rosenthal et al., 1970), have very low APase specific activities. These findings show that the differentiation of embryonal carcinoma is

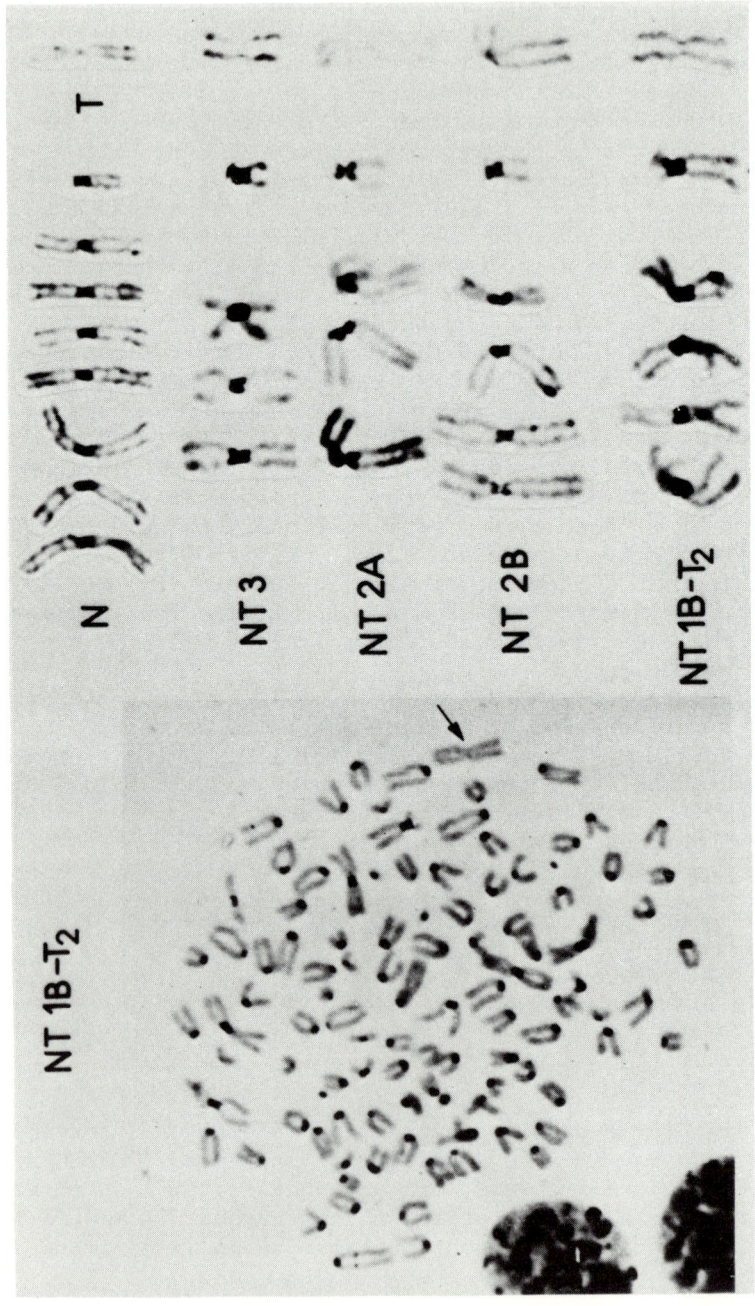

accompanied by a sharp decline in APase activity. Thus, at least with respect to this biochemical characteristic, embryonal carcinoma resembles normal differentiating embryonic cells. It is logical to hypothesize that the physiological significance of elevated APase activity in embryonal carcinoma is the same as in normal embryos. In this regard, the possibility of selecting variants for APase expression in tissue culture might provide information on the function of the enzyme that would be unattainable by using embryos.

Although we have demonstrated differences in heat denaturation, inhibition by L-phenylalanine and chromatographic properties between the embryonal carcinoma APase and a variety of tissue APases, the interpretation of these results is difficult due to the complex nature of the enzyme, which is a membrane-associated glycoprotein (references in Fernley, 1971 and Fishman, 1974). In particular, the fact that these experiments were performed on crude extracts or slightly purified APases reduces their rigor. The embryonal carcinoma APase has been purified about two hundred-fold in low yield (E.G.B., unpublished results). Further work along these lines should permit a better evaluation of the question posed: does the APase activity in embryonal carcinoma reflect expression of a gene(s) that is active in the adult or in the embryo?

It has been our experience that murine APases are poorly resolved in a variety of electrophoretic systems. This is in part due to aggregation. Preliminary results (E.G.B., unpublished) show that APases from embryonal carcinoma and placenta (the only sources so far examined) retain activity in 0.1% SDS and that SDS-polyacrylamide gel electrophoresis appears to be a promising analytical method by which to compare these APases. It should be possible to obtain enough material from early embryos for analysis by this sensitive

---

Figure 4. C-banding of metaphase chromosomes in parental and hybrid cell lines. C-bands were revealed on standard metaphase preparations exactly as described by Dev et al. (1973). For some preparations, slides were pretreated for one hour at room temperature in 0.2 N HCl as described by Sumner (1972). (Left) C-banding of a metaphase of hybrid clone NT1B-T2. The H3.6 metacentric chromosome is indicated by the arrow. (Right) Metacentric chromosomes of N18TG2 (N), H3.6 (T) and four hybrid cell lines, each of which contains the H3.6 marker chromosome, devoid of centromeric heterochromatin. The total number of metacentrics shown for the hybrids is low in each case because of overlaps in metaphases with very large total numbers of chromosomes.

TABLE 4

ALKALINE PHOSPHATASE ACTIVITY IN PARENTAL AND HYBRID CELL LINES

Cell Line	Number of Determinations	Specific Activity of APase (U/mg) ± S.E.[a]
N18TG2	10	2.0 ± 0.2
H3.6	19	37.1 ± 1.5
NT1	17	42.0 ± 7.0
NT1A	17	69.1 ± 8.5
NT1B	7	63.2 ± 9.8
NT1D	4	87.8 ± 12.0
NT1E	9	99.6 ± 22.2
NT1F	4	226.8 ± 18.6
NT2A	6	283.2 ± 35.0
NT2B	5	119.5 ± 23.2
NT2E	4	183.8 ± 65.6
NT3	5	164.7 ± 24.4

[a]Standard error

technique.

The enhancement of APase activity in the NT hybrids is an unusual and potentially interesting phenomenon. Two reports of increases of enzyme activities in hybrid cells have been published. Minna et al. (1972) found that three of sixteen intraspecific hybrids between a line of neuroblastoma (not N18TG2) and L cells expressed about twice as much AChE activ-

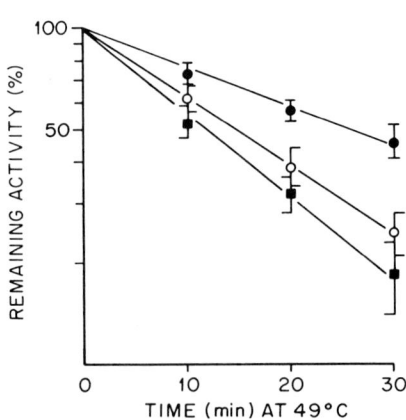

ity as the neuroblastoma parent, which expressed relatively low levels of the enzyme. Benedict et al. (1972) noted that in a cross between mouse fibroblasts and rat hepatoma cells the inducibility of aryl hydrocarbon hydroxylase by benz-[α]-anthracene, a characteristic of the fibroblast parent, was increased by factors of two to twenty in fifteen of sixteen hybrid cell lines examined. Uninduced levels of the enzyme were also elevated, but not always in proportion to the fully induced level (E. B. Thompson, personal communication).

The first question to be approached concerning the enhancement of APase activity must be that of which APase is being expressed in the NT hybrids. Our heat denaturation data suggest that it is the embryonal carcinoma enzyme, but further examination by other methods is required before this can be established with certainty. Improved electrophoresis or monospecific antibody to the teratoma APase should resolve this point.

The mechanism of enhancement of APase activity is unknown. It is, however, unlikely to be a dosage effect because the chromosomal contribution of N18TG2 (Table 3) is far greater than that of H3.6. The trivial explanation that the hybrids resulted from the fusion of N18TG2 with a subclass of H3.6 very rich in APase activity is unlikely since all of the original hybrid clones show enhancement, and no clones of embryonal carcinoma that have specific activities of APase approaching those observed in most hybrid lines have been isolated (Bernstine et al., 1973). Hybrids between H3.6 and a rat hepatoma (M. L. Hooper and B. E., unpublished results) express the very low level of APase activity (specific activity about 1.0) characteristic of the hepatoma cell line. Thus, enhancement of APase activity in hybrids of embryonal carcinoma appears not to be a general occurrence.

Cox et al. (1971) have presented evidence that the hormonal induction of APase in HeLa cells occurs by an enhancement of the catalytic efficiency of the enzyme that results in an increase in V_{max}. These workers propose that this change is mediated by a modifier molecule whose appearance requires RNA and protein synthesis. H. Koyama (unpublished

---

Figure 5. Heat inactivation of APases from parental and hybrid lines. Experiments were carried out as described by Bernstine et al. (1973). Samples were removed for assay at the times indicated. Error bars represent standard deviations based on five to eight determinations. ●—●, *N18TG2;* ○—○, *H3.6;* □—□, *NT1. Similar results were obtained for eight other lines of NT hybrid cells.*

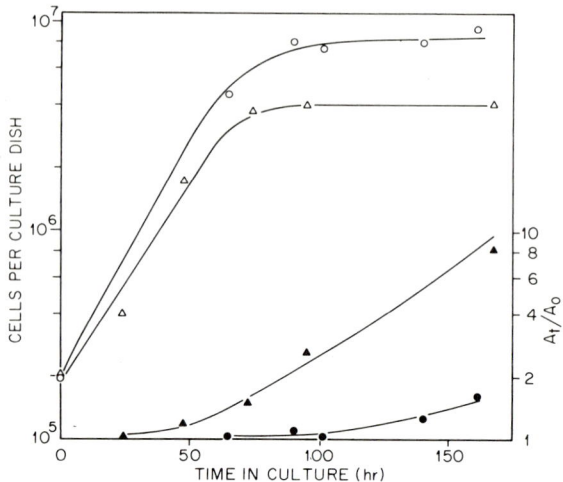

Figure 6. APase activity in N18TG2 and NT1 as a function of time in culture. At/Ao, ratio of specific activity at time t to specific activity at the initial time point. Cells per dish (o--o, N18TG2; △—△, NT1). At/Ao (●—●, N18TG2; ▲—▲, NT1).

results) has obtained data that show only slight differences between K_m and V_{max} of the APases of the parental lines and those of the hybrids NT2A and NT1E. This result makes models proposing a net increase in the catalytic efficiency of APase unlikely to explain the high activities in these lines.

The only presently available positive finding relevant to the APase levels in the hybrids is the large increase in activity noted during stationary phase (Fig. 6). It is of interest to note that AChE activity in N18TG2 increases by a factor of twenty-two (Minna et al., 1972) in stationary phase. Since both AChE and APase are associated with the plasma membrane, it is possible that research in the area of cell surface metabolism might provide some clues to explain these increases in activity.

The fact that we have observed no organized structures in tumors produced by the NT hybrids is consistent with the results obtained (Finch and Ephrussi, 1967; Jami et al., 1973) on hybrids between embryonal carcinoma and fibroblasts in which pluripotentiality is extinguished. It is noteworthy that a variety of cell types is found among the hybrids both in culture and in tumors. The biochemical characterization of these clones may give us some indication as to whether they express differentiated functions. The re-expression of

TABLE 5

ALKALINE PHOSPHATASE ACTIVITY IN SUBCLONES OF NT1B

Subclone	Specific Activity of APase (U/mg)[a]
1a	133.0
1b	0.10
4a	0.73
5	0.76
6	117.0
T1	8.4
T2	30.1

[a]Each value is the mean of two independent determinations.

pluripotentiality would be of particular interest. Clones NT1B-T1 and NT1B-T2 are morphologically embryonal carcinoma-like. Their future properties in culture and *in vivo* should be closely followed.

ACKNOWLEDGEMENTS

Most of the work reported here was carried out in collaboration with Drs. M. L. Hooper and H. Koyama. We thank Dr. Mary C. Weiss for many helpful discussions and for help with the karyological work. We received expert technical assistance from Mmes S. Grandchamp and N. Goussef, and Mlle. F. Ruiz.

Portions of the work were carried out during the tenure of fellowship to E.G.B. from the NSF-CNRS Exchange Program, the Pierre Philippe Foundation, and the L.A.V.F.W. This research was aided by a grant from the Centre National de la Recherche Scientifique (A.T.P. "Differenciation Cellulaire").

REFERENCES

Benedict, W. F., Nebert, D. W., Thompson, E. B. (1972). Expression of aryl hydrocarbon hydroxylase induction and suppression of tyrosine amino-transferase induction in somatic cell hybrids. Proc. Nat. Acad. Sci. U. S. 69, 2179.

Bernstine, E. G., Hooper, M. L., Grandchamp, S., Ephrussi, B. (1973). Alkaline phosphatase activity in mouse teratoma. Proc. Nat. Acad. Sci. U. S. 70, 3899.

Bernstine, E. G., Koyama, H., Ephrussi, B. (1975). The enhancement of alkaline phosphatase activity in hybrids between embryonal carcinoma and neuroblastoma cells. Proc. Nat. Acad. Sci. U. S., in press.

Cox, R. P., Elson, N. A., Shing-Hui, T., Griffin, M. J. (1971) Hormonal induction of alkaline phosphatase by an increase in catalytic efficiency of the enzyme. J. Mol. Biol. 58, 197.

Damjanov, I., Solter, D., Skreb, N. (1971). Enzyme histochemistry of experimental embryo-derived teratocarcinomas. Z. Krebsforsch. 76, 249.

Davidson, R. L. (1974). Control of expression of differentiated functions in somatic cell hybrids. In "Somatic Cell Hybridization" (Eds. R. L. Davidson and F. F. de la Cruz). Raven Press, New York, p. 131.

Dev, V. G., Miller, D. A., Miller, O. J. (1973). Chromosome markers in Mus musculus: strain differences in C-banding. Genetics 75, 663.

Ephrussi, B. (1972). "Hybridization of Somatic Cells." Princeton University Press, Princeton, New Jersey, p. 71.

Fernley, H. N. (1971). Mammalian alkaline phosphatases. In "The Enzymes" (Ed. P. D. Boyer). Academic Press, New York, p. 417.

Finch, B. W., Ephrussi, B. (1967). Retention of multiple potentialities by cells of a mouse testicular teratocarcinoma during prolonged culture *in vitro* and their extinction upon hybridization with cells of permanent lines. Proc. Nat. Acad. Sci. U. S. 57, 615.

Fishman, W. H. (1974). Perspectives on alkaline phosphatase isoenzymes. Amer. J. Med. 56, 617.

Jami, J., Failly, C., Ritz, E. (1973). Lack of expression of differentiation in mouse teratoma-fibroblast somatic cell hybrids. Exp. Cell Res. 76, 191.

Kahan, B., Ephrussi, B. (1970). Developmental potentialities of clonal *in vitro* cultures of mouse testicular teratoma. J. Nat. Cancer Inst. 44, 1015.

Kleinsmith, L. J., Pierce, G. B. (1961). Multipotentiality of single embryonal carcinoma cells. Cancer Res. 24, 1544.

Minna, J., Glazer, D., Nirenberg, M. (1972). Genetic dissection of neural properties using somatic cell hybrids. Nature New Biol. 235, 225.

Moog, F. (1965). Enzyme development in relation to functional differentiation. In "The Biochemistry of Animal Development" (Ed. R. Weber). Academic Press, New York, Vol. 1, p. 307.

Morton, R. K. (1953). Transferase activity of hydrolytic enzymes. Nature (London) 172, 65.

Mulnard, J. (1955). Contribution a la connaissance des enzymes dans l'ontogenese: les phosphomonoesterases acide et alkaline dans le developpement du rat et de la souris. Arch. Biol. 66, 525.

Pierce, G. B. (1967). Teratocarcinoma: model for a developmental concept of cancer. Curr. Topics Develop. Biol. 2, 223.

Rosenthal, M. D., Wishnow, R. M., Sato, G. H. (1970). *In vitro* growth and differentiation of clonal populations of multipotential mouse cells derived from a transplantable testicular teratocarcinoma. J. Nat. Cancer Inst. 44, 1001.

Stevens, L. C. (1970). The development of transplantable teratocarcinomas from intratesticular grafts of pre- and postimplantation mouse embryos. Develop. Biol. 21, 364.

Sumner, A. T. (1972). A simple technique for demonstrating centromeric heterochromatin. Exp. Cell Res. 75, 304.

THE RESPONSE OF MURINE TERATOCARCINOMA CELLS TO INFECTION WITH DNA AND RNA VIRUSES

JOHN M. LEHMAN, IRIS B. KLEIN AND ROSE M. HACKENBERG

Department of Pathology
University of Colorado Medical Center
Denver, Colorado   80220

 The murine teratocarcinoma described by Stevens and Little (1954) contains the highly malignant stem cell of the tumor, embryonal carcinoma, and differentiated tissues from all three germ layers (ectoderm, mesoderm, and endoderm). When this tumor is transplanted into the peritoneal cavity, free-floating aggregates of cells are produced (Pierce et al., 1960). These aggregates are called 'embryoid bodies' since they resemble 5 1/2-day-old mouse embryos with a center of embryonal carcinoma surrounded by a layer of well differentiated endoderm. Kleinsmith and Pierce (1964) made a single cell suspension of embryoid bodies and transplanted single cells into the peritoneal cavity. These experiments showed that a single embryonal carcinoma cell was able to produce a teratocarcinoma and was, therefore, multipotential.
 The somatic tissues that result from differentiation of embryonal carcinoma were shown to be benign by utilizing large cystic embryoid bodies. These embryoid bodies had mesenchyme, an inverted layer of endoderm, and in general did not contain embryonal carcinoma. When they were transplanted into strain 129 mice, the mesenchyme differentiated into bone, muscle, and cartilage, and the endoderm differentiated into glands and other types of epithelium. However, these tissues were benign and exhibited no growth potential (Pierce and Dixon, 1959), thus suggesting that these cells which were derived from the embryonal carcinoma were benign and that malignant cells could differentiate into benign cells.

The embryonal carcinoma of the testis is derived from the primordial germ cells, the precursor to germinal epithelium. This conclusion was drawn by histological and electron microscopic examination of early teratomas (Stevens, 1962; Pierce et al., 1967). Further confirmation for this hypothesis was obtained by the incorporation of a gene into the strain 129 mouse for the absence of germ cells which resulted in a marked reduction of spontaneous tumors (75% to 3%) (Stevens, 1967).

Pierce and Verney (1961), utlizing organ cultures, were successful in establishing the teratocarcinoma *in vitro*. Initially, the explants were invested by a layer of epithelium. The differentiated tissues of the explant died, leaving embryonal carcinoma cells, which, when transplanted into mice at 30 days in culture, produced typical teratocarcinomas. At 60 and 90 days, the tumors contained primarily muscle, suggesting that it might be possible to manipulate differentiation in these tumors. A number of other groups have described the establishment of monolayer cultures which retain the potential to differentiate *in vitro* (Kahan and Ephrussi, 1970; Rosenthal et al., 1970; Evans, 1972).

The embryonal carcinoma cell retains the potential for differentiation into representatives of all three germ layers. Preliminary experiments suggested that a tissue culture system could be developed in which stem cells could differentiate. With these facts in mind, we felt that this system might offer a model with which to study the effects of DNA or RNA viruses on cells that are in different stages of differentiation.

CELL CULTURE OF TERATOCARCINOMA

In our laboratory, we developed a cell culture model of the murine teratocarcinoma (Lehman et al., 1974). It was our expressed purpose to establish lines of the stem cell and to isolate lines of differentiated cells. The tumor used in these studies was OTT 6050, a subline of a trans-

---

ABBREVIATIONS AND TERMINOLOGY

FMF - flow microfluorometry.
Py - polyoma virus.
SV40 - Simian virus 40.
PFU - plaque-forming units.

For nomenclature of cell lines, see Table 1.

plantable teratocarcinoma kindly supplied by Dr. Leroy Stevens. Initial studies with the tumors showed that cells from embryoid bodies can grow in culture; within a few days to a week, complex patterns were observed in this monolayer culture system. In fact, it was obvious that in our routine subcultivation, we had at least four recognizable cell types: 1) embryonal carcinoma cells, 2) parietal yolk sac cells, 3) neuroepithelial-like cells, and 4) spindle cells which were smooth muscle or fibroblasts. If these cultures were fed but not subcultivated, a more complex pattern of differentiation was observed which included tissues such as cartilage, striated muscle, epithelial glands and squamous pearls. These results indicated that a multiplicity of differentiated tissues appeared in culture, similar to those observed in the transplantable tumor. This in turn suggested that the tissue culture model was extremely complex and might be difficult to utilize in studying particular cell types and their differentiation.

To alleviate this problem, we attempted to select for the stem cell. We were aided in our initial experiments by two factors: 1) the embryonal carcinoma grew as clumps, in colonies, which could be morphologically identified and 2) we were able to selectively remove these colonies by treatment with pancreatin (0.25% in phosphate buffered saline). The differentiated cells were more adherent to the surface; thus, these cells remained attached to the plastic surface while the embryonal carcinoma cells were easily removed. By frequent (every 24-48 hours) subcultivation, we were able to maintain a nearly pure population of embryonal carcinoma cells. If these cells were not subcultured frequently, differentiated cells appeared around the periphery of the colony.

The embryonal carcinoma cultures were further characterized in regard to 1) tumorigenicity, 2) morphology and 3) chromosome number distribution (Table 1). When individual colonies which contained between 20 and 500 cells were selected and injected in strain 129 mice, typical teratocarcinomas were obtained in 30% of the animals. The cells in these colonies were characterized morphologically by electron microscopy and found to be indistinguishable from embryonal carcinoma. These colonies were nearly pure embryonal carcinoma ($\sim$95%); differentiated cells, when present, appeared around the periphery of the colony. The third criterion used to define these cells was the chromosome number distribution and DNA content. Various passages of these cultures and embryoid bodies obtained from the

TABLE 1

CHARACTERISTICS OF CELL LINES ISOLATED FROM THE MURINE TERATOCARCINOMA

Cell Line	Morphology In Vitro	Tumors Produced Upon Transplantation	Chromosome Number (Stem Line)	DNA Content[b]
Embryonal Carcinoma	Colonies of small round cells[a]	Teratocarcinoma	Hypodiploid (39)	Diploid
Parietal Yolk Sac (PYS)	Epithelial	Parietal yolk sac carcinoma	Hypotetraploid	Near-tetraploid
Neuroepithelial 299-2 299-3	Stellate cells with processes	Neural tissues	Hypotetraploid (74-78)	Near-tetraploid
LTP (Mesenchymal)	Spindle cells	Undifferentiated fibrosarcoma	Hypotetraploid (74-78)	Near-tetraploid

[a] If not frequently subcultivated, differentiated cells appear, resulting in at least four morphologically distinct cells (embryonal carcinoma, PYS, neuroepithelium and spindle cells).

[b] DNA content measurements were performed by FMF. Normal mouse somatic cells served as standards.

transplantable tumor were fixed for metaphase chromosome analysis. All metaphase chromosome studies showed that a stem line number of 39 was present in 70-80% of the cells. A culture which has been maintained for three years still contains this stem line chromosome number.

We have also determined the DNA content (Horan et al., 1974) of the embryonal carcinoma cell line and other lines of differentiated cells obtained from the teratocarcinoma using the flow microfluorometry (FMF) system developed at the Los Alamos Scientific Laboratory. Utilizing the acriflavin-Feulgen staining procedure, single cells were analyzed for their DNA content. The cultures of embryonal carcinoma cells exhibited a diploid DNA content. When cultures of differentiated cells, such as parietal yolk sac (PYS) cells, were analyzed by FMF, these cells exhibited a near-tetraploid DNA content distribution. Chromosome analysis of the parietal yolk sac (PYS) cultures showed a hypotetraploid chromosome number distribution (74-78). At present we do not know the relationship of DNA content-chromosome number with the state of differentiation. However, the embryonal carcinoma cell lines have been near diploid in DNA content and chromosome number over the past three years suggesting the possibility that they may be used as a marker for the purity of the cultures of embryonal carcinoma. Studies on the relationship of the DNA content to neoplastic differentiation in this system are continuing with Dr. D. E. Swartzendruber utilizing FMF and cytophotometry. Preliminary results suggest a correlation between differentiation and shifts in DNA content. Whether this is a cause or effect will be defined more fully in further experiments.

On the basis of the above data, we have selected lines of embryonal carcinoma which can be propagated and retain the potential to differentiate. We have isolated and partially characterized a number of cell lines of differentiated cells derived from the embryonal carcinoma. In the majority of cases, attempts to selectively cultivate differentiated cells have been unsuccessful. Generally, they grow for a short period and then cease growing. However, we have been able to maintain three lines that are uniform in appearance and are not capable of producing teratocarcinomas. The PYS line is epithelial and makes basement membrane. We have selected a spindle cell (LTP) which will produce a mesenchymal tumor when injected into an animal. Two clones (299-2 and 299-3) which produce neuroepithelial tumors, are presently being characterized in regard to the production of specific neural enzymes. The characteristics of these lines and the embryonal carcinoma are listed in Table 1.

TABLE 2

RESPONSE OF MURINE TERATOCARCINOMA CELLS TO INFECTION WITH POLYOMA AND SV40

Cell Line	Polyoma			SV40		
	Presence of T Antigen	Virus Production	Transformation	Presence of T Antigen	Virus Production	Transformation
Embryonal Carcinoma	−	−	−	−	−	−
Mixed Teratocarcinoma[a]	±[b]	+	−	±[b]	−	±[c]
Parietal Yolk Sac (PYS)	±[d]	±	−	+	−	+
LTP	+	+	−	+	−	+

[a] These cultures contain embryonal carcinoma and differentiated cells.

[b] The cells that are positive are the differentiated cells.

[c] Lines of T antigen positive cells can be selected which have transformed cell characteristics. These cells are derived from the differentiated cell population. The embryonal cells are not infected.

[d] Two cell lines were assayed. One was infected and one was refractory.

INFECTION OF TERATOCARCINOMA CELLS WITH SV40 AND POLYOMA VIRUS

Utilizing the conditions mentioned above, we are able to selectively cultivate embryonal carcinoma which differentiates spontaneously. With the availability of this cell line and various differentiated cell lines, we have been able to compare the response of teratocarcinoma cells at various stages of differentiation to infection with DNA viruses.

The rationale for studying the effects of viruses in teratocarcinoma cells was to determine whether viral expression could be regulated by the cells at different states of differentiation. Initially, experiments were done with two oncogenic DNA viruses, polyoma (Py) and Simian virus 40 (SV40). These two papovaviruses offer the advantage of being small, with few gene products, and the expression of viral information is partially characterized. We had the availability of an early marker, the expression of viral specific tumor (T) antigen, and a late marker, the production of viral capsid protein (Defendi and Lehman, 1966; Lehman and Defendi, 1970). Virus replication could also be assayed by the production of viral DNA and replication of infectious particles.

Preliminary experiments using T antigen production as the marker for infection showed that both viruses were capable of infecting the differentiated cell lines (PYS and LTP), but were not capable of infecting the stem cell embryonal carcinoma (Swartzendruber and Lehman, 1975). The response of the differentiated cells was similar to the response of mouse somatic cells to Py or SV40 infection (Table 2). The differentiated cells were permissive for Py, causing complete cell lysis with virus replication. However, the differentiated cells were nonpermissive to complete SV40 replication. Following infection with SV40, transformed lines were obtained which showed an increase in saturation density and ability to grow in methyl cellulose (Lehman et al., 1975). When mixed cultures of teratocarcinoma (embryonal carcinoma and differentiated cells) were infected with SV40 virus, the percentage of T antigen-positive cells approached 95-98% but never 100%. In fact, it was possible to show by cloning, cytological and immunohistochemical techniques, that the negative cells were embryonal carcinoma. These cells were capable of producing teratocarcinomas when injected into strain 129 mice and no virus could be rescued by cell fusion. One clone of embryonal carcinoma was isolated from these infected mixed cultures and passaged for a number of generations. Both the embryonal carcinoma and its differentiated progeny were not producing the SV40 T antigen when assayed by immunofluorescence, showing that if the genome was present, it was not capable of expression.

As mentioned above, Py lysed the differentiated cells yielding high titers of Py. A carrier state was obtained since newly produced differentiated cells were infected and subsequently lysed, liberating virus. The embryonal carcinoma replicated, producing both stem cells and susceptible differentiated cells (**Table 2**). These cultures were maintained for four months and shed approximately 10^9-10^{11} plaque forming units (PFU) per culture. When cloned from the infected cultures, neither the embryonal carcinoma cells nor their differentiated cells produced the Py-T antigen.

These studies suggest that the response of teratocarcinoma cells to infection with two oncogenic DNA viruses was dependent on the state of differentiation of the cells. Numerous examples are available in which embryonic cells respond to virus infection differently than do their adult counterparts. An understanding of these responses to virus infection, dependent upon the state of differentiation, would be useful clinically and possibly offer new approaches to therapy.

The mechanism for this block to infection of embryonal carcinoma cells with Py and SV40 was studied by analyzing the early events in virus infection such as adsorption, penetration and infection with naked DNA (Swartzendruber and Lehman, in preparation). The adsorption studies were performed with SV40 virions that were labeled with ^{3}H-thymidine. These studies showed that there were no differences in adsorption between the stem cell and the differentiated progeny. Therefore, the block to infection was not at the initial step of adsorption. Further studies utilizing cell fractionation and electron microscopy exhibited the penetration of the virion into the cytoplasm and nucleus. These experiments confirmed that penetration followed the adsorption step normally. Uncoating of the virion was bypassed by utilizing infectious DNA and monitoring for the presence of the viral-specific T antigen. The differentiated cells from teratocarcinoma and other permissive and nonpermissive cells (mouse, human, and monkey) were infected with the viral DNA. However, the embryonal carcinoma with up to 30 µg of DNA per culture (1-5 x 10^6 cells/culture) did not exhibit any cells expressing the SV40 T antigen. The infectious activity of the DNA was 10^4 PFU per µg of DNA. This is an interesting result since numerous studies have shown that while virions may be blocked in the adsorption step of infection due to lack of viral receptor sites on the cell surface, infectious

DNA has been used successfully in infecting these cells. At present, we do not have an explanation for this result; possibly these stem cells may have a unique system to inactivate or degrade this incoming DNA. The fact that embryonal carcinoma is derived from the primordial germ cell suggests that the germ line may have developed a unique method to conserve its genetic information from foreign DNA. We plan to pursue this problem by probing the infected cell with DNA-DNA or RNA-DNA hybridization to detect the presence of the complete or partial viral genome in the embryonal carcinoma cell. Further work will also involve the analysis of possible enzyme systems in the embryonal carcinoma cell which are efficient in degrading the virion and viral DNA.

There are a number of systems which show that embryonic cells respond to virus infection differently than do their differentiated counterparts. Two such model systems have been defined for Py. Dawe et al. (1966) showed that with mouse submandibular glands, Py oncogenesis involves a complex interaction of the epithelio-mesenchymal complex. The neoplastic response to Py requires a certain amount of morphogenesis to have occurred and this response is then retained throughout adult life unless the epiethelio-mesenchymal relationships are disturbed. The second system involves the mouse kidney and its response to Py infection (Saxen et al., 1962). These studies show that the uninduced metanephrogenic mesenchyme is susceptible to Py infection, but the differentiated cells failed to support Py replication. Early mouse embryos are susceptible to infection with SV40 virus (Biczysko et al., 1973) and integrated copies of the viral genome can be detected in adult tissues of mice infected as blastocysts (Jaenisch and Mintz, 1974). These studies suggest that there are basic differences in the response of embryonic cells to viral infection. These studies are of practical value when considering that many viral infections are clinically important in regard to congenital abnormalities. The teratocarcinoma system may offer a model system which will define the relationship of virus infection in embryonic and adult cells.

INFECTION OF TERATOCARCINOMA CELLS WITH AN RNA VIRUS, MENGOVIRUS

Having defined the response of the teratocarcinoma cells to infection with two DNA viruses, we wished to study the response of this cell system to an RNA virus. For these studies we selected the picornavirus, mengovirus. This virus is small (27-28 mµ), containing a single-stranded RNA of 2×10^6 daltons which is approximately 30% of the particle. Virus synthesis begins approximately 2.5-3 hours postinfection and reaches a maximum in 6-8 hours. Lysis of the cell occurs in approximately 12 hours.

In the studies outlined below, we utilized three cell types: 1) mouse L cells, 2) PYS, and 3) embryonal carcinoma. The L cell provided the cell system for virus production and a plaque assay system for comparison of the viral growth in the teratocarcinoma cells. The teratocarcinoma cells, PYS and embryonal carcinoma, and L cells were seeded into 60 mm petri dishes. After 2-3 days the cells were infected with 5 PFU/cell of mengovirus. At 0, 3, 6, 9 and 12 hours postinfection, the cells and supernatant were harvested. These preparations were plaque assayed on L cell monolayers (Brownstein and Graham, 1961). The results of these experiments are shown in Table 3. An increase in virus titer is detected between 3 and 6 hours in all cells. Both PYS and L cells maximally produce approximately 700-1000 PFU/cell. However, the embryonal carcinoma cells produced between 80-140 PFU/cell. This is approximately 20% of virus production

TABLE 3

MENGOVIRUS PRODUCTION IN VARIOUS TERATOCARCINOMA CELL LINES[a]

Cells	Hours Postinfection				
	0	3	6	9	12
Embryonal Carcinoma	1.4[b]	0.7	80	140	70
Parietal Yolk Sac (PYS)	0.7	1.5	500	700	800
L Cells	1.2	1.2	600	500	1000

[a] Cells were infected with 5 PFU/cell and harvested at the times indicated.

[b] Virus production is expressed in terms of PFU/cell.

for PYS and L cells. Cytological examination of all cultures showed that by 8-12 hours postinfection all cells were lysed. At present we do not have an explanation for the results with embryonal carcinoma cells; however, it is evident that this small RNA virus is capable of replication and lysis of the teratocarcinoma cells, suggesting that at least with mengovirus the mechanism that blocks infection with Py and SV40 is not operative. Gwatkin (1963) reported that the zona pellicuda did not protect mouse ova from infection with mengovirus. When comparisons were made as to the virus produced, two facts emerged: 1) the growth cycle was longer for the ova when compared to L cells and mouse embryo fibroblasts and 2) even though the virus produced per ovum was greater, it was not what would be expected when the size of the cells were compared (Gwatkin, 1966; Gwatkin and Auerbach, 1966). On a unit volume comparison, the egg produces 1-2% of the virus yield of the L cells and mouse embryo fibroblasts. No information is available to explain this small yield of virus.

While our results do not show a lengthening of the viral replication cycle, there is a reduction in PFU of virus produced by embryonal carcinoma cells. Further analysis of viral replication and viral protein synthesis are underway in the laboratory in our effort to explain the mechanism for the decrease in replication of mengovirus in embryonal carcinoma cells.

CONCLUDING REMARKS

The murine teratocarcinoma displays neoplastic differentiation which may be useful in defining problems related to differentiation and cancer. We have developed a number of cell lines from the teratocarcinoma stem cell and differentiated cells, which we have characterized in regard to growth potential, cytogenetics, tumorigenicity and markers of differentiation. These cell lines have been utilized to study the response to virus infection of cells in different states of differentiation. When these cell lines were infected with two DNA viruses, SV40 and Py, the embryonal carcinoma cell lines showed no evidence of viral expression; however, the differentiated cells responded to these viruses, as do mouse somatic cells. Studies with the DNA viruses suggest that the block to the expression of viral T antigen was not an early infection step (adsorption or penetration), but a later step involved with early transcription or translation. An RNA virus, mengovirus, infected the embryonal carcinoma, but the virus yield was markedly reduced. This cell system may be useful in defining the cell-virus interactions which may be dependent upon the state of differentiation. Thus, we are continuing our studies of virus-cell interactions with this unique cell system.

ACKNOWLEDGEMENTS

This work was supported in part by grants CA-16030, CA-13419 and CA-15823 from the National Cancer Institute and a grant from the Milheim Foundation.

REFERENCES

Biczysko, W., Solter, D., Pienkowski, M., Koprowski, H. (1973). Interactions of early mouse embryos with oncogenic viruses, Simian virus 40 and polyoma. I. Ultrastructural studies. J. Nat. Cancer Inst. 51, 1945.

Brownstein, B., Graham, A. F. (1961). Interaction of Mengovirus with L cells. Virology 14, 303.

Dawe, C. J., Morgan, W. D., Slatich, M. J. (1966). Influence of epithelio-mesenchymal interactions on tumor induction by polyoma virus. Int. J. Cancer 1, 419.

Defendi, V., Lehman, J. M. (1965). Transformation of hamster embryo cells *in vitro* by polyoma virus: Morphological, karyological, immunological and transplantation characteristics. J. Cell. Comp. Physiol. 66, 315.

Evans, M. J. (1972). The isolation and properties of a clonal tissue strain of pluripotent mouse teratoma cell. J. Embryol. Exp. Morph. 28, 163.

Gwatkin, R. B. L. (1963). Effect of viruses on early mammalian development. I. Action of Mengo encephalitis virus on mouse ova cultivated *in vitro*. Proc. Nat. Acad. Sci. U.S. 50, 576.

Gwatkin, R. B. L. (1966). Effect of viruses on early mammalian development. III. Further studies concerning the interaction of Mengo encephalitis virus with mouse ova. Fertil. Steril. 17, 411.

Gwatkin, R. B. L., Auerbach, S. (1966). Synthesis of a ribonucleic acid virus by the mammalian embryo. Nature (London) 209, 993.

Horan, P. K., Jett, J. H., Romaro, A., Lehman, J. M. (1974). Flow microfluorometry analysis of DNA content in Chinese hamster cells following infection with Simian virus 40. Int. J. Cancer 14, 514.

Jaenisch, R., Mintz, B. (1974). Simian virus 40 DNA sequences in DNA of healthy adult mice derived from preimplantation blastocysts injected with viral DNA. Proc. Nat. Acad. Sci. U.S. 71, 1250.

Kahan, B. W., Ephrussi, B. (1970). Developmental potentialities of clonal *in vitro* cultures of mouse testicular teratoma. J. Nat. Cancer Inst. 44, 1015.

Kleinsmith, L. D., Pierce, G. B. (1964). Multipotentiality of single embryonal carcinoma cells. Cancer Res. 24, 1544.

Lehman, J. M., Defendi, V. (1970). Changes in deoxyribonucleic acid synthesis regulation in Chinese hamster cells infected with Simian virus 40. J. Virol. 6, 738.

Lehman, J. M., Speers, W. C., Swartzendruber, D. E., Pierce, G. B. (1974). Neoplastic differentiation: Characteristics of cell lines derived from a murine teratocarcinoma. J. Cell Physiol. 84, 13.

Lehman, J. M., Speers, W. C., Swartzendruber, D. E. (1975). Differentiative and virologic studies of teratocarcinoma *in vitro*. Intern. Congress Ser. 349, 176.

Pierce, G. B., Dixon, F. J. (1959). Testicular teratomas. I. Demonstration of teratogenesis by metamorphosis of multipotential cells. Cancer 12, 573.

Pierce, G. B., Dixon, F. S., Verney, E. L. (1960). Teratogenic and tissue forming potentials of the cell types comprising neoplastic embryonal bodies. Lab. Invest. 9, 583.

Pierce, G. B., Verney, E. L. (1961). An *in vitro* and *in vivo* study of differentiation in teratocarcinomas. Cancer 14, 1017.

Pierce, G. B., Stevens, L. C., Nakane, P. K. (1967). Ultrastructural analysis of early development of teratocarcinomas. J. Nat. Cancer Inst. 39, 755.

Rosenthal, M. C., Wishnow, R. M., Sato, G. H. (1970). *In vitro* growth and differentiation of clonal populations of multipotential mouse cells derived from a transplantable testicular teratocarcinoma. J. Nat. Cancer Inst. 44, 1001.

Saxen, L., Vainio, T., Toivonen, S. (1962). Effect of polyoma virus on mouse kidney rudiments *in vitro*. J. Nat. Cancer Inst. 29, 597.

Stevens, L. C., Little, C. C. (1954). Spontaneous testicular teratomas in an inbred strain of mice. Proc. Nat. Acad. Sci. U.S. 40, 1080.

Stevens, L. C. (1962). Testicular teratomas in fetal mice. J. Nat. Cancer Inst. 28, 247.

Stevens, L. C. (1967). Origin of testicular teratomas from primordial germ cells in mice. J. Nat. Cancer Inst. 38, 549.

Swartzendruber, D. E., Lehman, J. M. (1975). Neoplastic differentiation: Interaction of Simian virus 40 and polyoma virus with murine teratocarcinoma cells *in vitro*. J. Cell Physiol. 85, 179.

VIII.
Bibliography

BIBLIOGRAPHY

This comprehensive bibliography of experimental and spontaneous teratomas in rodents also includes selected references related to the same or similar tumors in other mammals including man, and in birds and plants. We apologize to all authors whose work might have been inadvertently omitted. We are grateful to Dr. Ivan Damjanov for preparing references concerning human teratomas.

Aldrich, J. T., Stevens, L. C. (1967). Effect of 5-fluorouracil on early teratomas in mice. Cancer Res. 27, 945-949.
Altman, R. P., Randolph, J. G., Lilly, J. R. (1974). Sacrococcygeal teratoma: American Academy of Pediatrics Surgical Section Survey-1973. J. Pediat. Surg. 9, 389-398.
Artzt, K., Bennett, D. (1972). A genetically caused embryonal ectodermal tumor in the mouse. J. Nat. Cancer Inst. 48, 141-158.
Artzt, K., Bennett, D., Jacob, F. (1974). Primitive teratocarcinoma cells express a differentiation antigen specified by a gene at the T-locus in the mouse. Proc. Nat. Acad. Sci. U. S. 71, 811-814.
Artzt, K., Dubois, P., Bennett, D., Condamine, H., Babinet, C., Jacob, F. (1973). Surface antigens common to mouse cleavage embryos and primitive teratocarcinoma cells in culture. Proc. Nat. Acad. Sci. U. S. 70, 2988-2992.
Artzt, K., Jacob, F. (1974). Absence of serologically detectable H-2 on primitive teratocarcinoma cells in culture. Transplantation 17, 632-634.
Askanazy, M. (1907). Die Teratome nach ihrem Bau, ihrem Verlauf, ihrer Genese und im Vergleich zum experimentellen Teratoid. Verh. Dtsch. Ges. Path. 11, 39-82.
Askanazy, M. (1909). Die Resultate der experimentellen Forschung uber teratoide Geschwulste. Wien. Med. Wschr. 59, 2518-2578.
Atkin, N. B. (1973). High chromosome numbers of seminomata and malignant teratoma of the testis: a review of data on 103 tumours. Brit. J. Cancer 28, 275-279.
Auerbach, R. (1972). The use of tumors in the analysis of inductive tissue interactions. Develop. Biol. 28, 304-309.

BIBLIOGRAPHY

Auerbach, R. (1972). Controlled differentiation of teratoma cells. In "Cell Differentiation" (Eds. R. Harris, P. Alin and D. Viza) Munksgaard, Copenhagen, pp. 119-123.

Bagg, H. J. (1936). Experimental production of teratoma testis in the fowl. Amer. J. Cancer 26, 69-84.

Beck, J. S. Fulmer, H. F., Lee, S. T. (1969). Solid malignant ovarian teratoma with "embryoid bodies" and trophoblast differentiation. J. Path. 99, 67-73.

Bernstine, E. G., Hooper, M. L., Grandchamp, S., Ephrussi, B. (1973). Alkaline phosphatase activity in mouse teratoma. Proc. Nat. Acad. Sci. U. S. 70, 3899-3903.

Berry, C. L., Keeling, J., Hilton, C. (1969). Teratoma in infancy and childhood: a review of 91 cases. J. Path. 98, 241-252.

Billington, W. D., Graham, C. F., McLaren, A. (1968). Extra-uterine development of mouse blastocysts cultured *in vitro* from early cleavage stages. J. Embryol. Exp. Morph. 20, 391-400.

Boon, T., Buckingham, M. E., Dexter, D. L., Jakob, H., Jacob, F. (1974). Teratocarcinome de la souris: isolement et proprietes de deux lignee de myoblastes. Ann. Microbiol. (Inst. Pasteur) 125B, 13-28.

Braun, A. C. (1959). A demonstration of the recovery of the crown-gall tumor cell with the use of complex tumors of single-cell origin. Proc. Nat. Acad. Sci. U. S. 45, 932-938.

Braun, A. C. (1969). "The Cancer Problem. A Critical Analysis and Modern Synthesis". Columbia University Press, New York and London.

Braun, A. C. (1972). The relevance of plant tumor systems to an understanding of the basic cellular mechanisms underlying tumorigenesis. Progr. Exp. Tumor Res. 15, 165-187.

Bresler, V. M. (1959). Experimental teratoids of white mouse testis induced by testosterone and copper sulfate. (In Russian) Vop. Oncol. 5, 24-30.

Bresler, V. M. (1964). On the dynamics of blastomogenesis in the testis. Acta Un. Int. Cancr. 20, 1501-1503.

Brinster, R. L. (1974). The effect of cells transferred into the mouse blastocyst on subsequent development. J. Exp. Med. 140, 1049-1056.

Buc-Caron, M.-H. Gachelin, G., Hofnung, M., Jacob, F. (1974). Presence of mouse embryonic antigen on human spermatozoa. Proc. Nat. Acad. Sci. U. S. 71, 1730-1733.

Caruso, P. A., Marsh, M. R., Minkowitz, S., Karten, G. (1971). An intense clinicopathologic study of 305 teratomas of the ovary. Cancer 27, 343-348.

Climie, A. R. W., Heath, L. P. (1968). Malignant degeneration

of benign cystic teratomas of the ovary. Review of the
literature and report of a chondrosarcoma and carcinoid
tumor. Cancer 22, 824-832.

Collins, D. H., Pugh, R. C. B. (1964). Classification and
frequency of testicular tumours. Brit. J. Urol. 36,
suppl. 2, 1-11.

Corfman, P. A. Richart, R. M. (1964). Chromosome number and
morphology of benign ovarian cystic teratomas. New Eng.
J. Med. 271, 1241-1244.

Damjanov, I., Solter, D. (1973). Yolk sac carcinoma grown
from explanted mouse egg cylinder. Arch. Path. 95,
182-184.

Damjanov, I., Solter, D. (1974). Experimental teratoma.
Curr. Topics Path. 59, 69-130.

Damjanov, I., Solter, D. (1974). Host-related factors determine the outgrowth of teratocarcinomas from mouse egg-cylinders. Z. Krebsforsch. Klin. Onkol. 81, 63-69.

Damjanov, I., Solter, D. (1974). Embryo-derived teratocarcinomas elicit splenomegaly in syngeneic host. Nature
(London) 249, 569-571.

Damjanov, I., Solter, D., Belicza, M., Skreb, N. (1971).
Teratomas obtained through extrauterine growth of seven-day old mouse embryos. J. Nat. Cancer Inst. 46, 471-480.

Damjanov, I., Solter, D., Serman, D. (1973). Teratocarcinoma
with the capacity for differentiation restricted to
neuro-ectodermal tissue. Virchows Arch. Abt. B 13,
179-195.

Damjanov, I., Solter, D., Skreb, N. (1971). Enzyme histo-chemistry of experimental embryo-derived teratocarcinomas. Z. Krebsforsch. 76, 249-256.

Damjanov, I., Solter, D., Skreb, N. (1971). Teratocarcinogenesis as related to the age of embryos grafted under
the kidney capsule. Wilhelm Roux' Arch. 167, 288-290.

Dehner, L. P. (1973). Intrarenal teratoma occurring in
infancy: report of a case with discussion of extra-gonadal germ cell tumors in infancy. J. Pediat. Surg.
8, 369-378.

Dexter, D. L., Buc-Caron, M.-H., Jakob, H., Nicolas, J.,
Gachelin, G. (1974). Teratocarcinome de la souris:
etude des antigenes de la surface des cellules a potentialites multiples. Ann. Microbiol. (Inst. Pasteur)
125B, 347-356.

Dixon, F. J., Moore, R. A. (1952). "Tumors of the Male Sex
Organs". Armed Forces Institute of Pathology, Washington, D. C.

Dunn, G. R., Stevens, L. C. (1970). Determination of sex of
teratomas derived from early mouse embryos. J. Nat.

Cancer Inst. 44, 99-105.

Dunn, T. B., Andervont, H. B. (1963). Histology of some neoplasms and non-neoplastic lesions found in wild mice maintained under laboratory conditions. J. Nat. Cancer Inst. 31, 873-901.

Edidin, M., Gooding, L. R., Johnson, M. (1974). Surface antigens of normal early embryos and a tumour model system useful for their further study. In "Immunological Approaches for Fertility Control" (Ed. E. Diczfalusy). Karolinska Institutet, Stockholm, pp. 336-356.

Edidin, M., Patthey, H. L., McGuire, E. J., Sheffield, W. D. (1971). An antiserum to "embryoid body" tumor cells that reacts with normal mouse embryos. In "Embryonic and Fetal Antigens in Cancer". (Eds. N. G. Anderson and J. H. Coggin, Jr.). National Technical Information Service, Springfield, Virginia, pp. 239-248.

Elgort, D. A., Abelev, G. I., Levina, D. M. Marienbach, E. V., Martochkina, G. A., Laskina, A. V., Solovjeva, E. A. (1973). Immunoradioautography test for alpha-fetoprotein in the differential diagnosis of germinogenic tumours of the testis and in the evaluation of effectiveness of their treatment. Int. J. Cancer 11, 586-594.

Evans, M. J. (1972). The isolation and properties of a clonal tissue culture strain of pluripotent mouse teratoma cells. J. Embryol. Exp. Morph. 28, 163-173.

Failly-Crepin, C., Uriel, J. (1973). A fetal carboxylic esterase of mouse origin common to normal embryonic brain and murine experimental teratocarcinoma. FEBS Letters, 33, 351-354.

Falin, L. I. (1940). Experimental teratoma testis in the fowl. Amer. J. Cancer 38, 199-211.

Fauve, R. M., Hevin, B., Jakob, H., Gaillard, J. A., Jacob, F. (1974). Antiinflamatory effects of murine malignant cells. Proc. Nat. Acad. Sci. U. S. 71, 4052-4056.

Fauve, R. M., Jakob, H., Gaillard, J. A., Jacob, F. (1974). Proliferation et differenciation de cellules primitives de teratocarcinome chez des souris allogenes traitees par une endotoxine. C. R. Acad. Sci. Paris, 278D, 1919-1922.

Fawcett, D. W. (1950). Bilateral ovarian teratomas in a mouse. Cancer Res. 10, 705-707.

Fawcett, D. W. (1950). The development of mouse ova under the capsule of the kindey. Anat. Rec. 108, 71-91.

Fawcett, D. W., Wislocki, G. B., Waldo, C. M. (1947). The development of mouse ova in the anterior chamber of the eye and in the abdominal cavity. Amer. J. Anat. 81,

413-443.
Fekete, E., Ferrigno, M. A. (1952). Studies on a transplantable teratoma of the mouse. Cancer Res. 12, 438-440.
Fellous, M., Gachelin, G., Buc-Caron, M.-H., Dubois, P., Jacob, F. (1974). Similar location of an early embryonic antigen on mouse and human spermatozoa. Develop. Biol. 41, 331-337.
Finch, B. W., Ephrussi, B. (1967). Retention of multiple developmental potentialities by cells of a mouse testicular teratocarcinoma during prolonged culture *in vitro* and their extinction upon hybridiztion with cells of permanent lines. Proc. Nat. Acad. Sci. U. S. 57, 615-621.
Fraumeni, J. F. Jr., LI, F. P., Dalager, N. (1973). Teratomas in children: epidemiologic features. J. Nat. Cancer Inst. 51, 1425-1430.
Gaillard, J. A. (1974). Differentiation and organization in teratomas. In "Neoplasia and Cell Differentiation" (Ed. G. V. Sherbet). S. Karger, Basel, pp. 319-349.
Gardner, R. L., Johnson, M. H. (1972). An investigation of inner cell mass and trophoblast tissues following their isolation from the mouse blastocyst. J. Embryol. Exp. Morph. 28, 279-312.
Gearhart, J. D., Mintz, B. (1974). Contact-mediated myogenesis and increased acetylcholinesterase activity in primary cultures of mouse teratocarcinoma cells. Proc. Nat. Acad. Sci. U. S. 71, 1734-1738.
Giovanella, B. C., Stehlin, J. S., Williams, L. J., Jr. (1974). Heterotransplantation of human malignant tumors in "nude" thymusless mice II. Malignant tumors induced by injection of cell cultures derived from human solid tumors. J. Nat. Cancer Inst. 52, 921-930.
Gooding, L. R., Edidin, M. (1974). Cell surface antigens of a mouse testicular teratoma. Identification of an antigen physically associated with H-2 antigens on tumor cells. J. Exp. Med. 140, 61-78.
Grobstein, C. (1950). Production of intra-ocular hemorrhage by mouse trophoblast. J. Exp. Zool. 114, 359-374.
Grobstein, C. (1950). Behavior of the mouse embryonic shield in plasma clot culture. J. Exp. Zool. 115, 297-314.
Grobstein, C. (1951). Intra-ocular growth and differentiation of the mouse embryonic shield implanted directly and following *in vitro* cultivation. J. Exp. Zool. 116, 501-526.
Grobstein, C. (1952). Intra-ocular growth and differentiation of clusters of mouse embryonic shields cultured with and

without primitive entoderm and in the presence of possible inductors. J. Exp. Zool. 119, 355-380.
Grobstein, C. (1952). Effect of fragmentation of mouse embryonic shields on their differentiative behavior after culturing. J. Exp. Zool. 120, 437-456.
Guenet, J. L., Jakob, H., Nicolas, J.-F., Jacob, F. (1974). Teratocarcinome de la souris: etude cytogenetique de cellules a potentialities multiples. Ann. Microbiol. (Inst. Pasteur) 125A, 135-151.
Guthrie, J.: Attempts to produce seminomata in the albino rat by inoculation of hydrocarbons and other carcinogens into normally situated and ectopic testes. Brit. J. Cancer 10, 134-144 (1956).
Guthrie, J. (1964). Observations on the zinc induced testicular teratomas in fowl. Brit. J. Cancer 18, 130-142.
Guthrie, J. (1971). Zinc induction of teratomas in Japanese quail (Coturnix Coturnix japonica) after photoperiodic stimulation of testis. Brit. J. Cancer 25, 311-314.
Guthrie, J., Guthrie, O. A. (1974). Embryonal carcinomas in Syrian hamsters after intratesticular inoculation of zinc chloride during seasonal testicular growth. Cancer Res. 34, 2612-2614.
Hsu, Y.-C., Baskar, J. (1974). Differentiation *in vitro* of normal mouse embryos and mouse embryonal carcinoma. J. Nat. Cancer Inst. 53, 177-185.
Jackson, E. B., Brues, A. M. (1941). Studies on a transplantable embryoma of a mouse. Cancer Res. 1, 494-498.
Jakob, H., Boon, T., Gaillard, J., Nicolas, J.-F., Jacob, F. (1973). Teratocarcinome de la souris: Isolement, culture et proprietes de cellules a potentialites multiple. Ann. Microbiol. (Inst. Pasteur) 124B, 269-282.
Jakupcevic, M., Lackovic, A., Damjanov, I., Bulat, M. (1974). Biogenic amines in retransplantable neurogenic teratocarcinoma. Experientia 30, 652-653.
Jami, J., Failly, C., Ritz, E. (1973). Lack of expression of differentiation in mouse teratoma-fibroblast somatic cell hybrids. Exp. Cell Res. 76, 191-199.
Jami, J., Ritz, E. (1973). Expression of tumor-specific antigens in mouse somatic cell hybrids. Cancer Res. 33, 2524-2528.
Jami, J., Ritz, E. (1974). Multipotentiality of single cells of transplantable teratocarcinomas derived from mouse embryo grafts. J. Nat. Cancer Inst. 52, 1547-1552.
Javadpour, N., Thomas, W., Bush, I. M. (1971). Production and treatment of testicular teratocarcinoma induced by genital ridge transplantation. J. Urol. 105, 534-539.
Jollie, W. P. (1961). The incidence of experimentally pro-

duced abdominal implantations in the rat. Anat. Rec. 141, 159-167.

Kahan, B. W., Ephrussi, B. (1970). Developmental potentialities of clonal *in vitro* culture of mouse testicular teratomas. J. Nat. Cancer Inst. 44, 1015-1036.

Kahan, B. W., Levine, L. (1971). The occurrence of a serum fetal α_1 protein in developing mice and murine hepatomas and teratomas. Cancer Res. 31, 930-936.

Kirby, D. R. S. (1963). The development of mouse blastocysts transplanted to the scrotal and cryptorchid testis. J. Anat. 97, 119-130.

Kirby, D. R. S. (1970). The extra-uterine mouse egg as an experimental model. Advan. Biosciences 4, 255-273.

Kleinsmith, L. J., Pierce, G. B., Jr. (1964). Multipotentiality of single embryonal carcinoma cells. Cancer Res. 24, 1544-1552.

Lehman, J. M., Speers, W. C., Swartzendruber, D. E., Pierce, G. B. (1974). Neoplastic differentiation: Characteristics of cell lines derived from a murine teratocarcinoma. J. Cell. Physiol. 84, 13-28.

Levak-Svajger, B., Skreb, B. (1965). Intraocular differentiation of rat egg cylinders. J. Embryol. Exp. Morph. 13, 243-253.

Levak-Svajger, B., Svajger, A. (1971). Differentiation of endodermal tissues in homografts of primitive ectoderm from two-layered rat embryonic shields. Experientia 27, 683-684.

Levak-Svajger, B., Svajger, A. (1974). Investigation on the origin of the definitive endoderm in the rat embryo. J. Embryol. Exp. Morph. 32, 445-459.

Levak-Svajger, B., Svajger, A., Skreb, N. (1969). Separation of germ layers in presomite rat embryos. Experientia 25, 1311-1312.

Levine, A. J., Torosian, M., Sarokhan, A. J., Teresky, A. K. (1974). Biochemical criteria for the *in vitro* differentiation of embryoid bodies produced by a transplantable teratoma of mice. The production of acetylcholine esterase and creative phosphokinase by teratoma cells. J. Cell. Physiol. 84, 311-318.

Linder, D., Hecht, F., McCaw, B. K., Campbell, J. R. (1975). Origin of extragonadal teratomas and endodermal sinus tumours. Nature (London) 254, 597-598.

Linder, D., McCaw, B. K., Hecht, F. (1975). Parthenogenetic origin of benign ovarian teratomas. New Eng. J. Med. 292, 63-66.

Linder, D., Power, J. (1970). Futher evidence for postmeiotic origin of teratomas in the human female. Ann. Human Genet. 34, 21-30.

Markert, C. L. (1968). Neoplasia: a disease of cell differentiation. Cancer Res. 28, 1908-1914.

Martin, G. R. (1975). Teratocarcinomas as a model system for the study of embryogenesis and neoplasia: a review. Cell 5, 229-243.

Martin, G. R., Evans, M. J. (1974). The morphology and growth of a pluripotent teratocarcinoma cell line and its derivatieves in tissue culture. Cell 2, 163-172.

Martin, G. R., Evans, M. J. (1975). Differentiation of clonal lines of teratocarcinoma cells: formation of embryoid bodies *in vitro*. Proc. Nat. Acad. Sci. U. S. 72, 1441-1445.

McGowan, L., Davis, R. H., Bunnang, B. (1971). Peritoneal fluid cytology associated with a solid ovarian teratoma in a mouse. Acta Cytol. 15, 306-309.

McLaren, A., Tarkowski, A. K. (1963). Implantation of mouse eggs in the peritoneal cavity. J. Reprod. Fert. 6, 385-392.

Meier, H., Myers, D. D., Fox, R. R., Laird, C. W. (1970). Occurrence, pathological features, and propagation of gonadal teratomas in inbred mice and in rabbits. Cancer Res. 30, 30-34.

Meins, F., Jr. (1972). Stability of the tumor phenotype in crown gall tumors of tobacco. Prog. Exp. Tumor Res. 15, 93-109.

Michalowsky, J. (1926). Die experimentelle Erzeugung einer teratoiden Neubildung der Hoden beim Hahn, Cbl. Allg. Path. Path. Anat. 38, 585-587.

Moore, M. A. S., Metcalf, M. (1970). Ontogeny of the haemopoietic system: yolk sac origin of *in vivo* and *in vitro* colony forming cells in the developing mouse embryo. Brit. J. Haematol. 18, 279-299.

Mosinger, M. (1961). Sur la carcinoresistance du cobaye. I. Les tumeurs spontanees du cobaye. Bull. Ass. Franc. Cancer, 48, 217-235.

Mostofi, F. K., Price, E. B. (1973). "Tumors of the Male Genital System." Armed Forces Institute of Pathology, Washington, D. C.

Mount, B. M., Stevens, L. C. (1971). The early natural history of murine germinal testicular tumors. J. Urol. 105, 812-816.

Mount, B. M., Stevens, L. C., Whitmore, W. F., Jr. (1970). The effect of chemotherapy on germinal testicular tumors in mice. Cancer 26, 570-576.

Nakahara, W., Tokuzen, R., Fukuoka, F. (1967). A transplantable hyalinogenic tumor of the mouse. Gann 58, 475-478.

Nefzger, M. D., Mostofi, F. K. (1972). Survival after surgery for germinal malignancies of the testis. I. Rates of survival in tumor groups. Cancer 30, 1225-1232.
Nefzger, M. D., Mostofi, F. K. (1972). Survival after surgery for germinal malignancies of the testis. II. Effects of surgery and radiation therapy. Cancer 30, 1233-1240.
Nicolas, J.-F., Dubois, P., Jakob, H., Gaillard, J., Jacob, F. (1975). Teratocarcinome de la souris: differenciation en culture d'une lignee de cellules primitives a potentialites multiples. Ann. Microbiol. (Inst. Pasteur), 126A, 3-22.
Nicholas, J. S. (1934). Experiments on developing rats, I. Limits of foetal regeneration; behavior of embryonic material in abnormal environments. Anat. Rec. 58, 387-413.
Nicholas, J. S. (1942). Experiments on developing rats, IV. The growth and differentiation of eggs and egg-cylinders when transplanted under the kidney capsule. J. Exp. Zool. 90, 41-71.
Oppenheimer, S. B., Edidin, M., Orr, C. W., Roseman, S. (1969). An L-glutamine requirement for intercellular adhesion. Proc. Nat. Acad. Sci. U. S. 63, 1395-1402.
Oppenheimer, S. B., Odencrantz, J. (1972). A qualitative assay for measuring cell agglutination: Agglutination of sea urchin embryo and mouse teratoma cells by concanavalin A. Exp. Cell Res. 73, 475-480.
Ozdzenski, W. (1969). Fate of primordial germ cells in the transplanted hind gut of mouse embryos. J. Embryol. Exp. Morph. 22, 505-510.
Pantoja, E. Noy, M. A. Axtmayer, R. W., Colon, F. E., Pelegrina, I. (1975). Ovarian dermoids and their complications. Comprehensive historical review. Obst.-Gynec. Surv. 30, 1-20.
Payne, J. M., Payne, S. (1961). Placental grafts in rats. J. Embryol. Exp. Morph. 9, 106-116.
Petrow, N. N. (1908). Experimentelle Embryonalimpfungen. Ein Beitrag zur Lehre von den Geschwulsten. Beitr. Path. Anat. 43, 1-42.
Peyron, A. (1939). Faits nouveaux relatifs a l'origine et a l'histogenese des embryomes. Bull. Ass. Franc. Cancer 28, 658-681.
Pierce, G. B., Jr. (1961). Teratocarcinomas, a problem in developmental biology. Canad. Cancer Conf. 4, 119-137.
Pierce, G. B. Jr. (1966). Ultrastructure of human testicular tumors. Cancer 19, 1963-1983.
Pierce, G. B. (1967). Teratocarcinoma: model for a devel-

opmental concept of cancer. Curr. Topics Develop. Biol. 2, 223-246.

Pierce, G. B. (1970). Differentiation of normal and malignant cells. Fed. Proc. 29, 1248-1254.

Pierce, G. B. (1972). Differentiation and cancer. In "Cell Differentiation" (Eds. R. Harris, P. Alin and D. Viza) Munksgaard, Copenhagen, pp. 109-114.

Pierce, G. B. (1974). Neoplasms differentiations and mutations. Amer. J. Path. 77, 103-118.

Pierce, G. B. (1974). The benign cells of malignant tumors. In "Developmental Aspects of Carcinogenesis and Immunity" (Ed. T. J. King). Academic Press, New York, pp. 3-22.

Pierce, G. B., Abell, M. R. (1970). Embryonal carcinoma of the testis. Pathol. Annu. 5, 27-60.

Pierce, G. B., Beals, T. F. (1964). The ultrastructure of primordial germinal cells and fetal testes and of embryonal carcinoma cells in mice. Cancer Res. 24, 1553-1567.

Pierce, G. B. Bullock, W. K., Hungtington, R. W., Jr. (1970). Yolk sac tumors of the testis. Cancer 25, 644-658.

Pierce, G. B. Dixon, F. J. (1959). Testicular teratomas. I. Demonstration of teratogenesis by metamorphosis of multipotential cells. Cancer 12, 573-583.

Pierce, G. B., Dixon, F. J. (1959). Testicular teratomas. II. Teratocarcinoma as an ascitic tumor. Cancer 12, 584-589.

Pierce, G. B., Dixon, F. J., Verney, E. L. (1958). The biology of testicular cancer. II. Endocrinology of transplanted tumors. Cancer Res. 18, 204-206.

Pierce, G. B., Dixon, F. J., Verney, E. L. (1960). Teratocarcinogenic and tissue forming potentialities of the cell types comprising neoplastic embryoid bodies. Lab. Invest. 9, 583-602.

Pierce, G. B., Johnson, L. D. (1971). Differentiation and cancer. *In Vitro* 7, 140-145.

Pierce, G. B., Midgley, A. R. Jr. Sri Ram J., Feldman, J. D. (1962). Parietal yolk sac carcinoma: clue to the histogenesis of Reichert's membrane of the mouse embryo. Amer. J. Path. 41, 549-566.

Pierce, G. B., Stevens, L. C., Nakane, P. K. (1967). Ultrastructural analysis of the early development of teratocarcinoma. J. Nat. Cancer Inst. 39, 755-773.

Pierce, G. B., Verney, E. L. (1961). *In vitro* and *in vivo* study of differentiation in teratocarcinomas. Cancer 14, 1017-1029.

Pierce, G. B. Jr., Verney, E. L., Dixon, F. J. (1957). The biology of testicular cancer. I. Behavior after trans-

plantation. Cancer Res. 17, 134-138.
Pierce, G. B., Wallace, C. (1971). Differentiation of malignant to benign cells. Cancer Res. 31, 127-134.
Porter, D. C. (1967). Observations on the development of mouse blastocysts transferred to the testis and kidney. Amer. J. Anat. 121, 73-86.
Pugh, R. C. B., Smith, J. P. (1964). Teratoma. Brit. J. Urol. 36, suppl. 2, 28-44.
Riviere, M., Chouroulinkov, J., Guerin, M. (1960). Production de tumeurs par injections intratesticulaires de chlorure de zinc chez le rat. Bull. Ass. Franc. Cancer. 47, 57-87.
Robboy, S. J., Scully, R. E. (1970). Ovarian teratoma with glial inplants on the peritoneum. An analysis of 12 cases. Hum. Pathol. 1, 643-653.
Rosenthal, M. D., Wishnow, R. M., Sato, G. H. (1970). *In vitro* growth and differentiation of clonal populations of multipotential mouse cells derived from a transplantable testicular teratocarcinoma. J. Nat. Cancer Inst. 44, 1001-1014.
Runner, M. N. (1947). Development of mouse eggs in the anterior chamber of the eye. Anat. Rec. 98, 1-13.
Salaun, J. (1964). Sur la formation experimentale d' embryome chez le rat et le poulet. Arch. Anat. Micr. Morph. Exp. 53, 387-396.
Salaun, J. (1968). Sur la formation de teratomes chez le poulet et le rat. Arch. Anat. Micr. Morph. Exp. 57, 11-34.
Saltykow, S. (1900). Uber Transplantation zusammengesetzter Teile. Wilhelm Roux' Arch. 9, 329-409.
Sherman, M. I. (1975). The culture of cells derived from mouse blastocysts. Cell 5, 343-349.
Simard, L. C. (1957). Polyembryonic embryoma of the ovary of parthenogenetic origin. Cancer 10, 215-223.
Skakkebaek, N. E. (1972). Possible carcinoma in-situ of the testis. Lancet 2, 516-517.
Skakkebaek, N. E. (1975). Atypical germ cells in the adjacent 'normal' tissue of testicular tumours. Acta Path. Microbiol. Scand. A 83, 127-130.
Skreb, N., Damjanov, I., Solter, D. (1972). Teratomas and teratocarcinomas derived from rodent egg-shields. In "Cell Differentiation." (Eds. R. Harris, P. Alin and D. Viza). Munksgaard, Copenhagen, pp. 151-155.
Skreb, N., Svajger, A. (1973). Histogenetic capacity of rat and mouse embryonic shields cultivated *in vitro*. Wilhelm Roux' Arch. 173, 228-234.

BIBLIOGRAPHY

Skreb, N., Svajger, A., Levak-Svajger, B. (1971). Growth and differentiation of rat egg-cylinders under the kidney capcule. J. Embryo. Exp. Morph. 25, 47-56.

Sobis, H., Vandeputte, M. (1974). Development of teratomas from displaced visceral yolk sac. Int. J. Cancer, 13, 444-453.

Solter, D., Damjanov, I., Koprowski, H. (1975). Embryo-derived teratoma: a model system in developmental and tumor biology. In "The Early Development of Mammals" (Eds. M. Balls and A. E. Wild). Cambridge University Press, London, pp. 243-264.

Solter, D., Skreb, N., Damjanov, I. (1970). Extrauterine growth of mouse egg-cylinders results in malignant teratoma. Nature (London) 227, 503-504.

Stevens, L. C. (1958). Studies on transplantable testicular teratomas of strain 129 mice. J. Nat. Cancer Inst. 20, 1257-1276.

Stevens, L. C. (1959). Embryology of testicular teratomas in strain 129 mice. J. Nat. Cancer Inst. 23, 1249-1295.

Stevens, L. C. (1960). Embryonic potency of embryoid bodies derived from a transplantable testicular teratoma of the mouse. Develop. Biol. 2, 285-297.

Stevens, L. C. (1962). Testicular teratomas in fetal mice. J. Nat. Cancer Inst. 28, 247-268.

Stevens, L. C. (1962). The biology of teratomas including evidence indicating their origin from primordial germ cell. Ann. Biol. 1, 585-610.

Stevens, L. C. (1964). Experimental production of testicular teratomas in mice. Proc. Nat. Acad. Sci. U. S. 52, 654-661.

Stevens, L. C. (1966). Development of resistance to teratocarcinogenesis by primordial germ cells in mice. J. Nat. Cancer Inst. 37, 859-861.

Stevens, L. C. (1967). The biology of teratomas. Advan. Morph. 6, 1-31.

Stevens, L. C. (1967). Origin of testicular teratomas from primordial germ cells in mice. J. Nat. Cancer Inst. 38, 549-552.

Stevens, L. C. (1968). The development of teratomas from intratesticular grafts of tubal mouse eggs. J. Embryol. Exp. Morph. 20, 329-341.

Stevens, L. C. (1970). Experimental production of testicular teratomas in mice of strains 129, A/He, and their F_1 hybrids. J. Nat. Cancer Inst. 44, 929-932.

Stevens, L. C. (1970). Environmental influences on experimental teratocarcinogenesis in testes of mice. J. Exp. Zool. 174, 407-414.

Stevens, L. C. (1970). The development of transplantable teratocarcinomas from intratesticular grafts of pre- and postimplantation mouse embryos. Develop. Biol. 21, 364-382.

Stevens, L. C. (1973). A new inbred subline of mice (129/ter Sv) with a high incidence of spontaneous congenital testicular teratomas. J. Nat. Cancer Inst. 50, 235-242.

Stevens, L. C. (1973). A developmental genetic approach to the study of teratocarcinogenesis. BioSci. 23, 169-172.

Stevens, L. C. Bunker, M. C. (1964). Karyotype and sex of primary testicular teratomas in mice. J. Nat. Cancer Inst. 33, 65-78.

Stevens, L. C., Hummel, K. P. (1957). A description of spontaneous congenital testicular teratomas in strain 129 mice. J. Nat. Cancer Inst. 18, 719-747.

Stevens, L. C., Little, C. C. (1954). Spontaneous testicular tumors in an inbred strain of mice. Proc. Nat. Acad. Sci. U. S. 40, 1080-1087.

Stevens, L. C. Mackensen, J. A. (1961). Genetic and environmental influences on teratocarcinogenesis in mice. J. Nat. Cancer Inst. 27, 443-453.

Stevens, L. C., Varnum, D. S. (1974). The development of teratomas from parthenogenetically activated ovarian mouse eggs. Develop. Biol. 37, 369-380.

Svajger, A., Levak-Svajger, B. (1974). Regional developmental capacities of the rat embryonic endoderm at the head-fold stage. J. Embryol. Exp. Morph. 32, 461-467.

Swartzendruber, D. E., Lehman, J. M. (1975). Neoplastic differentiation: Interaction of simian virus 40 and polyoma virus with murine teratocarcinoma cells *in vitro*. J. Cell. Physiol. 85, 179-188.

Teilum, G. (1971). "Special Tumors of Ovary and Testis". Lippincott, Philadelphia.

Teresky, A. K., Marsden, M., Kuff, E. L., Levine, A. J. (1974). Morphological criteria for the *in vitro* differentiation of embryoid bodies produced by a transplantable teratoma of mice. J. Cell. Physiol., 84, 319-332.

Thiery, M. (1963). Ovarian teratoma in the mouse. Brit. J. Cancer. 17, 231-234.

Toro, E. (1938). The homeogenetic induction of neural folds in rat embryos. J. Exp. Zool. 79, 213-236.

VandenBerg, S. R., Herman, M. M., Ludwin, S. K., Bignami, A. (1975). An experimental mouse testicular teratoma as a model for neuroepithelial neoplasia and differentiation. I. Light microscopic and tissue and organ culture observations. Amer. J. Path. 79, 147-168.

BIBLIOGRAPHY

Vink, H. H. (1970). Ovarian teratomas in guinea pigs: a report of ten cases. J. Path. Bact. 102, 180-182.
Willis, G. W., Hajdu, S. I. (1973). Histogically benign teratoid metastasis of testicular embryonal carcinoma: report of five cases. Amer. J. Clin. Pathol. 59, 338-343.
Willis, R. A. (1951). "Teratomas". Armed Forces Institute of Pathology, Washington, D. C.
Willis, R. A. (1962). "The Borderland of Embryology and Pathology", 2nd. ed. Butterworth, London.
Willis, R. A., Rudduck, H. B. (1943). Testicular teratomas in horses. J. Path. Bact. 55, 165-171.
Woodruff, J. D. Protos, P., Peterson, W. F. (1968). Ovarian teratomas. Relationship of histologic and ontogenic factors to prognosis. Amer. J. Obstet. Gynec. 102, 702-715.
Wurster, K., Hedinger, C., Mienberg, O. (1972). Orchioblastmartige Herde in Hodenteratomen von Erwachsenen. Virchows Arch. Pathol. Anat. 357, 231-242.

Subject Index

A

A9 cell line, 126, 129-133
Acetylcholine esterase, 75, 256-260, 282-284
AChE, *see* Acetylcholine esterase
Acid phosphatase, 199, 200
Aggregate, cell, 170, 174, 178, 182, 183, 192-203, 240, 241
Alkaline phosphatase, 177, 181, 199, 200, 213, 214, 217, 271-285
  DEAE cellulose fractionation, 274, 276
  heat denaturation, 273, 279, 281-283
  inhibition by L-phenylalanine, 273-274, 281
  intestinal, 273
  kidney, 273, 274, 276
  liver, 273
  placental, 274, 276, 281
Allophenic mouse, 63-72
ALS, *see* Antiserum, rabbit anti-mouse lymphocyte
Antibody
  anti-F9, 163-164
  anti-H-2, 110, 112
Antigen
  cell surface, 101-106, 109-119, 123-134, 252, 253, 262-267
  F9, 101-106, 124, 163-164
  H-2, 103, 105, 106, 110-115, 119
  non-H-2, 111
  transplantation, 109-112
  tumor, 294-296, 299
Antiserum
  anti-PTC, 101-105
  rabbit anti-mouse lymphocyte, 151-153
  rabbit anti-mouse thymocyte, 148-153
APase, *see* Alkaline phosphatase
ATS, *see* Antiserum, rabbit anti-mouse thymocyte
402 AX; tumor line, 109, 112-119
8-Azaguanine, 191, 197

B

Blastocyst, 13, 18-30, 34-43, 63, 64, 66, 67, 103-104, 111, 116-118, 190, 191, 203, *see also* Inner cell mass, Trophectoderm
  cell lines, *see* MB2, MB4
  cell lines, antigenicity of, 123-134
  colonization with teratoma cells, 51-56, 62-72

haploid, 34-43
Brain
  chromosomal proteins in, 223-230
  ultrastructure, 216

C

Capping, 115, 130-131
Cell fusion, 271-285
Chimerism, 56, see also Allophenic mouse
Choriocarcinoma, 4-6
Chromatin, 221-230
Chromosome
  analysis, 43-47, 162-163, 165, 197, 199, 202, 276-281, 291-293
  metacentric, 162, 199
Cleavage stage embryo, see Embryo, preimplantation
Cl ld cell line, 116, 118
Clone, cells derived from teratoma, 161-166, 172, 179-184, 196, 200-202
  efficiency, 200-201
Coat color, 53, 54, 56, 64-67, 71
Creatine phosphokinase, 75, 256, 258, 260

D

dif−, 163
DNA
  content of cells, 292-293
  virus, see Virus, DNA

E

Ectoderm, and ectodermal cell types, 13, 20-30, 89-90, 161, 163, 210-212, 214, 215, 243, 245
Egg, see Ovum
Egg-cylinder, 25, 27, 33, 42, 210-215, 217
Electrophoresis
  polyacrylamide gel, 216, 222-230, 257, 259, 281, 283
  starch gel, 65-67
Electron microscopy, 177, 179, 211-218, 291, 296
  scanning, 180, 253, 255
Embryo, see also Blastocyst, Egg-cylinder, Morula, Ovum, Parthenogenone, Parthenote
  postimplantation, 103, 104, 106, 116-118, 143
  postimplantation, teratomas induced from, 83-96, 144-156
  preimplantation, 109, 111, 116-119, 145, 153
  rat, 84, 85, 87-90, 95, 96
  susceptibility to viral infection, 297-299
Embryogenesis, relationships between teratoma development and, 17-30, 183-184, 189-191, 203
Embryoid body, 7, 17, 18, 21-30, 52, 60-64, 68, 70, 71, 73-77, 170, 172, 176-184, 192, 199, 200, 212, 217, 237, 240, 241, 244, 245, 247, 248, 251-266, 272-274, 289, 291
  simple, 170, 180, 182, 184, 241, 245, 248, 252, 272, 274
  cystic, 170, 180, 181, 182, 184, 241, 245, 248, 252, 253
Embryonal carcinoma, 4, 13, 14, 43, 47, 123, 131, 146, 169-179, 181-184, 189-192, 195-200, 202, 203, 222, 223, 238-241, 243-245, 248, 251-253, 260, 262-265, 267, 272-275, 279-281, 283-285, 289-299
Embryonic shield, rat, 84-87, 89-96

SUBJECT INDEX

Endoderm, and endodermal cell types, 14, 19-30, 89, 161, 163, 170, 177-179, 181-184, 210, 214, 239-243, 245-248, 252, 260
parietal, 217, see also Yolk sac, parietal
visceral, 217

F

F9 cell line, 123, 126, 128, 129, 133, see also Antibody, anti-F9 and Antigen, F9
Feeder layer, 171, 172, 175-178, 200
α-Fetoprotein, 76, 77
Fibroblast, 172, 190, 192, 197, 239, 241, 245, 252, 259, 266, 275, 283, 284, 291, 299
Filter, Nucleopore®, 192-194, 198-200
Flow microfluorometry, 292, 293

G

Gelatin substratum, 171, 174-177, 182
Genital ridge, 18, 142-144, 146, 154, 156
Germ cell, 5, 13, 53, 67, 72, see also Primordial germ cell
Germ layer, 4, 7, 13, 14, 68, 69, 90, 162, 163
separation of, 88-90
Glucose phosphate isomerase, 65-67
Graft
skin, 53-56, 112-114
tissue, 110-112

H

Hl, see Histone, lysine-rich
H-2, see Antigen, H-2
H3.6 cell lines, 272, 275-285
Head-fold stage embryo, 86, 87, 89-92, 94, 95
Hepatocyte, 65, 66, 68
Heterochromatin, centromeric, 277, 280, 281
Histology of cultured teratoma cells, 241-249
Histone, 221-227
H_o, 226
lysine-rich, 226
Hormone, 142, 144, 154, 156
Human chorionic gonadotrophin, 5, 6
Hybrid cell, 271, 274-285

I

ICM, see Inner cell mass
Immunfluorescence, 102-106, 112, 114-118, 127-134
Immunoperoxidase, 102-106
Infection, viral, of teratoma cells, 294-299
Inner cell mass, 13, 19, 20, 27, 29, 30, 52, 104, 106, 111, 116-118, 190, 191, 203

K

Karyotype, see Chromosome analysis
Keratin pearl, 241, 246, 247

L

L cell line, 230, 298, 299
Liver, chromosomal proteins in, 223-230
LS 420C, tumor line, 173, 239, 272, 273
LTP cell line, 292-295
Lymphocyte, 103, 110, 113, 114, *see also* Antiserum, rabbit anti-mouse lymphocyte

M

Major urinary protein complex, 65, 66
MB2 cell line, 124-126, 128, 129, 131
MB4 cell line, 124-134
Melanoblast, 65, 70
Melanocyte, 68
Mengovirus, 297-299
Mesenchyme, 7, 8, 245-248, 289, 292, 293, 297
Mesoderm and mesodermal cell types, 22, 23, 28, 29, 85, 89, 91, 95, 161, 163, 245, 248
Micromanipulation, 52
Monophyletic tumor, 215
Morula, 19-21, 63, 76-78, 101, 104, 116-118
Mouse strain, *see* Strain, mouse
Mutation, induction of, 161-166
Myogenesis, 73-75
Myoblast, 257
Myotube, 62, 74, 75

N

Neuroblastoma, 274-277
Neuroepithelium, 291-293
Neuron, 197, 240-243, 246, 247, 256, 257
NHC, *see* Protein, non-histone chromosomal
Nitrosoguanidine-induced mutations, 162-166
N18TG2 cell line, 272, 277, 278, 280-284

O

Organ culture, 92-95
OTT 5568 tumor line, 173, 238
OTT 6050 tumor line, 52, 102, 113, 114, 126, 128, 191, 199, 223, 253-255, 257, 258, 262-265, 272-274, 290
Ovarian teratoma, 17-30, 34, 47, 48, 141, 209-211
Ovariectomy, 151, 154, 155
Ovum, 116-118, *see also* Germ cell, oocyte, zygote

P

Parthenogenone, 33-48, *see also* Parthenote
 haploid, 33-48
 induction of, 34, 35
Parthenogenetic development
 induced, 36-39
 spontaneous, 19-30
Parthenote, 17-30, *see also* Parthenogenone
Parietal yolk sac, *see* Endoderm, parietal and yolk sac, parietal
PCC4.azal cell line, 126, 128, 129, 131, 162-165, 191-203, 265
PFU, *see* Plaque forming unit
Phosphatase, *see* Acid phosphatase, Alkaline phosphatase

SUBJECT INDEX

Pigment, see also Melanoblast, Melanocyte
  eye, 53
  producing cell, 240, 241, 245-247
Plaque-forming unit, 296, 298
Plasminogen, 259, 260, 266
Polyoma virus, 295-297, 299
Primitive streak, 21, 33, 28-30, 89, 90, 93, 95
Primordial germ cell, 3, 4, 9, 47, 73, 102, 209, 212
Protease, 259, 260, 266
Protein
  chromatin, 221-230
  non-histone chromosomal, 221, 222, 225-230
Py, see Polyoma
PYS, see Yolk sac, parietal
  cell line, 252, 264, 265, 292-294, 298

R

Reichert's membrane, 6, 14, 21, 24, 25, 28, 29, 124, 177, 179, 184, 243

S

Seminoma, 4, 5
SIKR cell line, 179, 238
Simian virus 40, 259, 266, 294-296, 299
Skin graft, 53-56, 112-114
Sl, see Steel gene
Sperm, 101, 104
  development, 104, 105
Spermatozoa, see Sperm
Spleen size, 147, 148, 150, 152, 155
Splenectomy, 148-150, 152

Splenomegaly, 148
Steel gene (*Sl*), 64, 65, 68, 71, 82, 102, 105, 140
Strain, mouse
  A, 146
  A/He, 77, 142-145
  AKR, 146, 147
  B10, 112-114
  CBA, 146
  C57Bl, 64-67, 72, 146-148, 151-155
  CF1, 116
  C3H, 34, 36, 38-42, 44, 102, 105, 146-150, 153, 154
  LG, 27
  LT, 18, 19, 26, 34, 141, 210
  129, 3, 18, 29, 34, 36, 38-42, 44-46, 53, 54, 56, 61, 64, 66-70, 72, 82, 101, 102, 109, 112-114, 123, 140-145, 162, 164, 173, 223, 251-254, 261, 263-266, 273, 277, 289, 291, 295
Strain, rat
  Fischer, 85, 87, 88
SV40, see Simian virus 40

T

3T3 cell line, 263, 266
T antigen, see Antigen, tumor
Terminology, recommendations, 13, 14
Testis, 103-105
Thymus, chromosomal proteins in, 223-230
Trophectoderm, 20, 104, 118
Trophoblast, 21, 22, 24, 25, 27, 40-43, 104, 106, 109, 116-118, 125, 190, 197, 198, 214
tum^-, 163-166
tum^+, 165-166

SUBJECT INDEX

U

Ultrastructure of teratoma, cells 209-217

V

Virus, *see also* Mengovirus, Polyoma, Simian virus 40
  endogenous, 213
  DNA, 289, 290, 295, 297, 299
  RNA, 289, 290, 297-299

Y

Y antigen, 56
Y chromosome, 43, 56, *see also* Chromosome analysis
Yolk sac, 43, 62, 74, 245, 253, *see also* PYS, cell line
  carcinoma, 4, 6, 8, 14, 77, 124, 252, 263
  parietal, 14, 104, 265, 291

Z

Zygote, 111, *see also* Germ cell, Ovum, Sperm